AF875574

HEURISTIC RISK MANAGEMENT

BE AWARE, GET PREPARED, DEFEND YOURSELF

MICHAEL LINES

Second Edition, updated digital edition published 260727.

ISBN: 978-1-964431-00-0 (ebook)

ISBN: 978-1-964431-01-7 (paperback)

ISBN: 978-1-964431-02-4 (hardcover)

ISBN: 978-1-964431-03-1 (audiobook)

Library of Congress Control Number: 2024909526

Cover design and photo from Canva.com

CONTENTS

ACKNOWLEDGMENTS

I am immensely grateful to my reviewers, Ken Crombie, JR Cunningham, Dave Farrow, Andrew Karpie, Barry Kortekaas, Greg Schaffer, and Glen Sorensen, for the feedback they provided regarding the book. Their insights, expertise and attention to detail enhanced its quality and clarity. Finally, I would like to extend my heartfelt thanks to all those who contributed in any capacity, no matter how small, to the completion of this project. Your contributions are deeply appreciated.

For my wife, Deborah, for her boundless patience and support of all my endeavors.

If you know the enemy and know yourself, you need not fear the result of a hundred battles. If you know yourself but not the enemy, for every victory gained, you will also suffer a defeat. If you know neither the enemy nor yourself, you will succumb in every battle.

— SUN TZU, THE ART OF WAR

INTRODUCTION

A heuristic technique is an approach to problem-solving that employs a practical method to achieve an answer that is not guaranteed to be optimal but sufficient for the intended purpose.

We are in a war, a cyber war. If you come away with anything from reading this book, I hope it will be the realization that every individual, business, organization, and nation is under constant attack from those who want to steal or harm their information, money, operations, or reputation. This book will teach you how to determine your most likely adversaries, prepare to defend against their attacks and evolve your plan and defenses based on real-world events.

Properly assessing and managing risk is the foundation of all effective information security programs. There will always be more threats, vulnerabilities, and issues than any organization has the time or resources to address. Prioritizing where to focus your limited attention and funding is critical to ensuring that these resources are directed toward delivering the most significant risk mitigation returns.

Yet despite the increasing focus on cyber risk assessment and management in legislation and control frameworks over the past decade, massive information breaches continue to be reported daily, often traceable to the failure to implement or adequately administer fundamental security controls.[1] Clearly, something is not working. Either risk assessment and management are not being done correctly, or they are not being done at all.

Risk assessment and management techniques do not need to be difficult to understand or implement; however, quantitative approaches such as Factor Analysis of Information Risk (FAIR)[2] and Operationally Critical Threat, Asset, and Vulnerability Evaluation (OCTAVE),[3] combined with government guidance such as in the U.S. National Institute of Standards and Technology (NIST) Risk Management Framework,[4] have overly complicated what should be a simple and intuitive process.

The U.S. government's default response to any significant cyber breach — imposing more regulations on organizations — is making the security leader's job increasingly challenging by burdening them with an ever-growing list of cybersecurity requirements.[5] The result is a compliance hamster wheel, with security teams constantly running to keep up with the latest regulations and, as a result, frequently failing to either implement or adequately maintain the fundamental security controls that matter the most in mitigating risk.

The Heuristic Risk Management (HRM) approach I propose in this book directly responds to these challenges. It is a straightforward and effective method that non-technical business leaders can easily understand, is simple to implement, and ensures that your efforts deliver tangible risk reduction returns.

The information security program and plan you develop following the HRM approach may be all your company needs if you are a small business and largely unregulated. However, even if your company is larger or heavily regulated, the foundation you develop following the HRM approach will help ensure that the controls you add over time enhance rather than dilute the effectiveness of those you already have in place.

Who Can Benefit From This Book?

This book was written for new security leaders in organizations with immature or nonexistent security functions. However, it can be just as helpful for security leaders with established programs to help perform a "sanity check" on their programs and ensure that they are driven by risk considerations and not compliance bureaucracy.

This book is also helpful for board members and senior leaders to educate them on what risk-driven information security programs and plans should address. If these leaders receive plans and updates that do not address the fundamental issues of threats and risk mitigation I cover, then it is time to ask your security and IT leaders more probing questions!

Small and Medium Businesses (SMBs) are particularly vulnerable targets for cybercriminals, a fact that should not be taken lightly. According to a survey conducted by CNBC and Momentive of small business owners,[6] almost half (42%) of the respondents admitted they have no formal plans for managing a security breach or incident. Given the challenging economic environment in which small businesses operate, it is understandable that they may struggle to allocate sufficient resources to cybersecurity. However, this vulnerability underscores the urgent need for effective security measures, regardless of the organization's size.

Even SMBs who know the risks and have the necessary funding still struggle due to a lack of cyber expertise or knowledge. In such cases, these organizations will often attempt to adopt a larger organization's compliance-focused approach. They will try to implement complex security control frameworks such as the U.S. National Institute of Standards and Technology (NIST) 800 series,[7] Federal Risk and Authorization Management Program (FedRAMP),[8] Payment Card Industry Data Security Standard (PCI DSS),[9] or the ISO/IEC 27000 standards[10] for information security management. Their (mistaken) belief is that implementing all the controls provided by a chosen framework will ensure the organization's security. However, it is essential to note that **compliance does not equal security**, as any experienced security professional will tell you, and as the continuing breaches at "compliant" organizations demonstrate.

What if your business must comply with a mandated control framework based on the industry in which you operate, such as PCI DSS for retailers accepting credit cards, the North American Electric Reliability Corporation Critical Infrastructure Protection (NERC CIP)[11] for the power system industry, or the Cybersecurity Maturity Model Certification (CMMC)[12] for U.S. defense contractors? The Heuristic Risk Management (HRM) approach is equally valuable in these instances.

By focusing on risk reduction rather than simply taking the cheapest or easiest actions, you are taking a prudent and cautious approach to enhancing your organization's security. The Heuristic Risk Management (HRM) approach will guide you in prioritizing the implementation of required controls for risk mitigation, after which you can focus on the remaining controls you need to demonstrate compliance with some standard. Following this approach will minimize your chances of being breached while you travel on your road to eventual compliance with the required regulations for your organization.

How This Book Is Organized

To understand how I have organized this book, you need to understand the concept of the OODA loop. The OODA loop, an acronym for Observe, Orient, Decide, and Act, was developed by military strategist and United States Air Force Colonel John Boyd.[13] Boyd conceived the OODA loop based on his experiences as a fighter pilot, where success in aerial combat heavily depended on making quick, accurate decisions in a rapidly changing environment. The core idea was that victory in a dogfight relies not just on the technical capabilities of the aircraft but significantly on the pilot's ability to observe the situation rapidly, orient themselves in the context of the enemy's movements, decide on the best course of action, and then act on that decision swiftly.

In cybersecurity, the principles of the OODA loop are not just applicable; they are crucial. Just as a fighter pilot must react quickly to a constantly changing battlefield, cybersecurity leaders face a digital battlefield perpetually in flux as attackers evolve their methods and targets. All the while, they must also deal with shifting internal business priorities due to the ever-changing business environment. Understanding the OODA loop is not just helpful; it is essential for cybersecurity survival.

- **Observe:** You must constantly monitor and assess the cybersecurity landscape, including threat intelligence feeds and business and cybersecurity news. Strategic risk management focuses on understanding and determining who and what you are working to protect your organization against. This understanding forms the basis for your cybersecurity program: "Why does your organization exist, and what purpose does it serve?". This aligns with the HRM tenet of "Be Aware" covered in Part One.
- **Orient:** You must then use the information gathered during the observation phase to clarify your security situation. This involves understanding your organization's unique environment, critical systems, business operations, and security controls—all the things that make up the constraints and opportunities in your environment. This is the first part of tactical risk management, which will be covered in Part Two.
- **Decide:** As a cybersecurity leader, you must decide on your best course of action based on the understanding you developed in the orientation phase. This might involve updating security policies, implementing new security measures, or reconfiguring existing defenses based on the organization's budget and business priorities. This is the second half of tactical risk management in Part Two, which will lead to finalizing your information security plan for protecting the organization. Orient and Decide together comprise the HRM tenet of "Get Prepared".

- **Act:** Finally, you must implement your plan, which is the focus of Operational Risk Management in Part Three. You must implement and adjust this plan based on the ever-shifting currents of the battle against your external enemies while appeasing your internal stakeholders. The lessons you learn in implementing your plan will feed into its evolution and the next iteration of the OODA loop. This forms the HRM tenet of "Defend Yourself" covered in Part Three.

In the dynamic world of cybersecurity, the OODA loop is not a one-time process but a continuous cycle. Just as a fighter pilot must navigate an ever-changing battlefield, as a cybersecurity leader, you must continuously cycle through the OODA loop, adapting to new information and changing circumstances. This continual process of observing, orienting, deciding, and acting enables cybersecurity teams to stay ahead of threats, much like a skilled pilot must outmaneuver and outthink their adversaries in the air if they are to both win and survive.

A Note On The Contents

If you are unfamiliar with risk terms and processes, you might want to read the chapter "Manage Your Risks" first, then jump back to the beginning to read the rest of the book. This chapter discusses the risk terms and calculations I use in this book, simplified to align with the heuristic approach I recommend. Even if you are familiar with risk terminology and processes, it is worth your time to skim this chapter so that you will understand where I am coming from when I use them, as this may differ from the more complex risk management processes you may be used to.

As you read the book, note that I have expressly not included a lot of detail regarding the topics I discuss but have instead provided footnotes to relevant articles and referenced sites if you would like to explore the topic further. I did this because I hate plowing through books that regurgitate details I already know, which I suspect was done only to inflate the page count to justify charging you more for the book. Refer to the referenced footnotes if you need to know more about a topic. If you don't, hopefully, you will find this "just the facts" format an easier read.

Instead of offering a detailed roadmap for every step in creating an information security program, this book encourages you to think about risk and security in a way that can be tailored to your specific needs. This approach will equip you with a practical, actionable roadmap to apply to your organization. The goal? To enhance your cybersecurity posture as swiftly as possible. Remember, your adversaries aren't waiting for you to get your act together![14]

In terms of terminology, I use the terms companies, businesses, and organizations interchangeably in this book. This is because cybersecurity threats and risks apply to all of these entities, from small retailers and nonprofits to large government agencies. Please don't take my use of these terms when discussing risk or security topics to mean that they only apply to that particular structure. Likewise, when I refer to the board (short for the board of directors), this can apply to the CEO and an equivalent management committee if this is the body with overall risk governance responsibility for the organization.

One final note: THIS IS NOT A COOKBOOK! My intention is NOT to provide you with a series of checklists that you can use to create and manage an information security program with all the requisite forms, slides, and examples. If anything, that approach is the problem I am attempting to overcome by teaching you to think about information security as a risk problem in a war context — not a compliance and checklist exercise. This book is about understanding and applying risk to designing, implementing, and managing an effective and evolving information security program.

PART I

STRATEGIC RISK MANAGEMENT

BE AWARE

Know your enemy, know his sword.

— MIYAMOTO MUSASHI, BOOK OF FIVE RINGS

1

KNOW YOUR ENEMY

> Water shapes its course according to the nature of the ground over which it flows; the soldier works out his victory concerning the foe he is facing.
>
> — SUN TZU, THE ART OF WAR

BEFORE YOU CAN PROTECT your organization from cyber threats, you need to know who and what you are fighting. Building your defenses without a clear picture of who your adversaries are is no different than swinging a sword in a dark room, hoping to evade your enemy's attacks. Yes, you can get lucky, but luck then is the measure of your program, not skill.

Part One, Strategic Risk Management (SRM), focuses on knowing your enemies, what they are after, and the threats their attacks pose to organizations. This information forms the foundation for your information security program, which encompasses the people, processes, and technologies you will deploy and manage to counter these threats. More importantly, it defines your strategic risks, which are based on the threats you are seeking to defend your organization against. Your strategic risks, once defined, should be relatively stable and usually only require review on an annual basis, as discussed in the chapter "Report Your Progress".

WHY DO you need to know who is attacking you? Isn't knowing their methods (phishing, malware, etc.) enough? "Aren't we wasting time?" you may ask, when attributing an attack to who is behind it is often impossible.

It matters because without knowing who your likely attackers are, you will not pay proper attention to news or reports concerning them. This is a key component of your threat intelligence efforts, a vital part of your information security organization that I will discuss in the chapter "Assemble Your Team". Knowing how your attackers evolve their attacks and techniques is critical to ensuring that your defenses are kept current to counter them.

Starting your information security program development by focusing on 'why' your cybersecurity program is needed rather than the traditional approach of focusing on 'what' controls are required will help you sell your program to the executive team. This approach allows you to plant the seeds you'll harvest when it's time to request the resources needed to implement your plan, a topic we'll cover in Part 2, "Tactical Risk Management".

In risk terminology, the entities you are attempting to defend your organization against are known as 'threat actors'.

These are the groups and individuals responsible for the threats to your business. In the world of cyber risk, there are four primary threat actors: nation-states, criminals and criminal organizations, hacktivists, and cyberterrorists.

1.1 Nation-States

Nation-state actors refer to groups that are directly or indirectly backed by or represent an individual nation's military or intelligence services. All major developed countries have offensive and defensive cyber warfare organizations and capabilities.

These agencies are tasked with carrying out defensive and offensive cyber operations to advance their national interests as directed by their country's leadership. Their activities can include everything from intelligence gathering[1] and defensive measures intended to protect the nation from attacks to offensive actions[2] against those deemed threats to the country. These offensive actions can range from digital-only attacks to cyberattacks supporting kinetic military actions.[3]

When it comes to attacks against most commercial enterprises, the focus of most nation-state actors is usually espionage (gathering of valuable intelligence on the enemy) or sabotage (disruption or destruction of the enemy's infrastructure or economic interests). However, some nation-states are also engaged in what are typically considered criminal activities[4] to help fund the nation and its military operations.

It is also important to note that some nation-state cyber players, Russia and China in particular,[5] have a strong relationship between their official cyber warfare agencies and quasi-independent criminal organizations operating within their borders.[6] These groups act as a proxy for attacks or perform intelligence-gathering operations, feeding information they discover back to the state intelligence agencies as part of their criminal activities.

There are currently four nation-state adversaries who primarily threaten U.S. interests: the People's Republic of China, Russia, North Korea and Iran.[7] Of these, China is by far the most active and represents the greatest danger to US businesses, individuals and government agencies.[8]

Some examples of nation-state attacks include:

- **OPM Data Breach (2015):** The United States Office of Personnel Management (OPM) suffered a massive data breach attributed to Chinese nation-state hackers.[9] This breach resulted in the theft of sensitive personal information of about 21.5 million people, including government employees and contractors. This incident remains one of the most significant data breaches of U.S. government data by a foreign government that has been publicized.
- **Equifax Breach (2017):** In one of the most significant data breaches ever, the personal information of approximately 147 million people was stolen from Equifax, a major credit bureau with operations worldwide. The U.S. Department of Justice charged four members of the Chinese military with the breach, indicating state-sponsored involvement.[10]
- **DPRK Supply Chain Attacks (2023):** DPRK (North Korea) state-linked cyber actors increasingly focus on software supply chain attacks to exploit organizations worldwide.[11] Supply chain attacks are highly effective in compromising numerous well-protected, high-profile targets.

- **Microsoft (2023):** Between May and June 2023, a threat actor known as Storm-0558, reportedly affiliated with the People's Republic of China, breached the Microsoft Exchange Online mailboxes of 22 organizations and over 500 individuals worldwide.[12] Utilizing authentication tokens signed by a Microsoft key established in 2016, the intruder gained unauthorized access to these accounts. Among those affected were senior officials from the United States government engaged in national security affairs.

The bottom line is that no company can consider itself immune from being attacked by nation-state attackers, either directly by a targeted attack or indirectly by malicious bots.[13]

1.2 Criminals

Criminal activity (directly attacking businesses for financial gain) can range from individual hackers and criminals to large-scale criminal enterprises organized to the same degree as commercial businesses. Criminal activities are almost always focused on making money via illegal means.

In contrast to the media image of hackers as lone wolves operating from a dark basement, cybercrime has become a multi-billion dollar global business and operates as such. Lone hackers can sell their services on dark web emporiums,[14] buy ready-made malware for purposes ranging from ransomware to credit card theft (with on-call technical support available),[15] rent botnets to conduct Distributed Denial of Service (DDOS) attacks against companies for extortion,[16] use money laundering services to help hide and deposit the money they have stolen,[17] etc.

Considering how many ways a criminal can steal money from a business using cyber means, the only way a company could consider itself immune from criminal attacks would be if it had no assets, bank accounts, computers, or staff. In that case, they would also not exist as a business!

Some prominent examples of attacks by criminal organizations include:

- **Target Corporation Breach (2013):** In one of the most infamous retail hacks, cybercriminals stole the credit and debit card information of approximately 40 million Target customers. The attackers infiltrated Target's network through a third-party vendor, showcasing the vulnerabilities in supply chain security.[18]
- **Global Banking Trojan Attacks:** Organized cybercrime groups have deployed banking Trojans like TrickBot and Emotet globally. These malware programs steal banking credentials and personal information, leading to substantial financial losses. For instance, the TrickBot Trojan has impacted financial institutions worldwide, causing millions of dollars in losses.[19]
- **Ransomware Attacks by Criminal Syndicates:** Cybercrime syndicates like LockBit have conducted widespread ransomware campaigns, encrypting organizational data and demanding hefty ransoms from organizations worldwide. These groups often target large corporations and governments, causing significant operational disruptions and costing

hundreds of millions in damages and ransom payments.[20]

Criminal Employees

Malicious employees are another class of criminal you need to consider.[21] While they may not have had criminal intent when you hired them, employees who "turn bad" can sabotage systems or processes to strike back at their employer because of perceived or actual harm to themselves. According to the 2022 Unit 42 Incident Response Report from Unit 42 by Palo Alto Networks, 75% of insider threat cases involved a disgruntled ex-employee who left with company data, destroyed company data, or accessed company networks after their departure.[22] Of course, employees can also have criminal intent from the start and join companies with the express intent of stealing information for personal gain. Examples of the damage that these types of attacks can cause include:

- **Tesla (2018):** A disgruntled Tesla employee made code changes to the company's Manufacturing Operating System and exported large amounts of highly sensitive Tesla data to unknown third parties.[23]
- **Elliott Greenleaf (2021):** Four law firm lawyers allegedly stole the organization's files and deleted its emails to help a competing law firm launch a new office.[24]
- **Georgia-Pacific (2014):** A fired employee caused more than $1.1 million in damages to the paper manufacturer's operations.[25]

- **Google (2023):** A Google employee was charged with stealing confidential information regarding Google's data center designs for large AI model development, that he used to found his AI start-up in China.[26]

1.3 Hacktivists

Hacktivist organizations, such as Anonymous,[27] use technology, hacking, and other techniques to bring about political or social change to support their cause.[28] Typically, they seek to gather sensitive information that can harm or pressure organizations to change positions or activities they oppose.

No organization is safe from such attacks, as demonstrated by these attacks against organizations of all types:

Government Organizations

- **Russia (2022):** In retaliation for the Russian invasion of Ukraine, Anonymous targeted many Russian government websites, including the state-controlled RT news.[29]
- **Black Lives Matter (2020):** In support of the Black Lives Matter movement, Anonymous is suspected of attacking the Minneapolis Police Department and the City of Minneapolis websites.[30]

Commercial Organizations

- **Ashley Madison Data Breach (2015):** The infidelity dating website Ashley Madison was targeted by a group called 'The Impact Team'. These hacktivists breached the site's security and released millions of user's personal information, citing the site's moral implications as the reason for the hack.[31]

- **Stratfor Email Leak (2011):** The global intelligence company Stratfor suffered a breach by Anonymous, resulting in the leak of thousands of emails and credit card details.[32]

Nonprofit Organizations

- **British Pregnancy Advisory Service (BPAS) Attack (2012):** The British Pregnancy Advisory Service, a UK-based charity providing abortion services, was targeted by a hacker who stole 10,000 database records. The individual defaced the website and threatened to release the details of women who had contacted the organization.[33]
- **Planned Parenthood (2015):** Planned Parenthood experienced a hacktivist attack after controversial videos were released that purported to show the organization's officials discussing the sale of fetal tissue. The attack involved a website defacement and the release of employee data.[34]

A hacktivist group can target any organization in the social spotlight because of its products or services, leadership, positions or statements, or the sensitive nature of the information it collects or processes. Nowadays, that can be just about any company, from the smallest retailer to the largest global corporation. Somehow, somewhere, there is somebody who is pissed off by what your company does or represents, and if they have the requisite technical skills, they can do something about it that can cause harm to your company.

1.4 Cyberterrorists

Cyberterrorism is a growing threat that blurs the lines between nation-state, criminal, and hacktivist attacks.[35] The primary differentiator is hacking and other cyber techniques that cause fear or harm to people or organizations and promote political, ideological, or religious aims. Cyberterrorists seek to achieve their objectives through violence or the threat of violence. They might launch attacks on critical infrastructure, such as power grids or transportation systems, or steal sensitive information to use it for criminal purposes to fund their operations.

An example of these groups' activities is the numerous attacks against critical infrastructure organizations across the U.S. attributed to hacking groups associated with the Iranian Government's Islamic Revolutionary Guard Corps. These attacks have targeted organizations using Israeli-made Unitronics Vision Series programmable logic controllers (PLC).[36] The attacks have ranged from defacing the controller's user interface to altering the operation of the systems that the PLC was controlling, potentially harming large numbers of people.[37]

Now that you know your enemies, rank them in the order in which you feel they are most likely to threaten your company. Different organizations — nonprofit, commercial, and governmental — attract different threat actors based on various factors, such as the nature of the data they hold, their public profile, and their strategic value. Knowing who will most likely attack you is vital to helping you prioritize your defenses effectively.

Nonprofit Organizations

The threat actors most likely to target nonprofit organizations are:

- **Hacktivists:** Nonprofit organizations, particularly those involved in contentious social, political, or environmental issues, are prime targets for hackers. These actors may seek to disrupt operations, steal data to further their cause or expose the nonprofit's activities.
- **Nation-State Actors:** In some cases, especially for nonprofits engaged in international aid or human rights advocacy, nation-state actors may target these organizations to gather intelligence or disrupt their activities.
- **Criminal Organizations and Individuals:** Many nonprofits have been attacked with ransomware, even though they may have limited funds to pay the ransom. This has significantly damaged their ability to fulfill their mission and, in some cases, caused them to cease operations.

Commercial Organizations

The following threat actors are most likely to target commercial organizations:

- **Criminal Organizations and Individuals:** Commercial entities, particularly those in the financial, healthcare, and retail sectors, are often targeted for financial gain through tactics like ransomware, data theft, and fraud.

- **Nation-State Actors:** Some industries, like technology, energy, or defense, may attract nation-state actors looking to steal intellectual property or disrupt critical infrastructure.

Governmental Agencies

The actors most likely to pose a threat to government agencies are:

- **Nation-State Actors:** Governmental agencies are primary targets for other nation-state actors seeking to gain strategic geopolitical advantages, conduct espionage, or disrupt national infrastructure.
- **Criminal Organizations and Individuals:** Ransomware attacks have increasingly targeted government agencies, from federal to local levels. However, these may be intended more to cause harm to their operations than to extort money.
- **Hacktivists:** Hacktivists who aim to protest against policies or expose governmental activities may also target governments.

There is no correct answer to the above prioritization; simply the answer that makes sense to you and your company's leadership when you use this information to present your program, as I will cover in the chapter "Present Your Plan" in Part Two.

2

KNOW YOURSELF

> If you know the enemy and know yourself, your victory will not stand in doubt.
>
> — SUN TZU, THE ART OF WAR

YOU MUST REALIZE you are in a war when dealing with cyber threats. Before you even walk through the door, your organization is under attack from the threat actors we just discussed in the prior chapter. Just as with an officer dropped into the middle of a firefight, your first step (if you hope to survive) had better be to seek cover and understand who is shooting at you and from where.[1] It should not be to inspect your soldier's uniforms or the grammar and punctuation in their after-action reports, at least if you hope to survive the day.

From an HRM perspective, "Know Yourself" involves two questions: what matters to your company's leaders, and what is it about what your organization has or does that is desirable to those likely to attack it? Getting answers to these questions is the focus of this chapter.

You first need to understand what matters most to your company's leaders. While that sounds simple and obvious, I have lost count of the number of security leaders I've encountered who do not have a clue as to their leaders' priorities. Instead, they are wrapped around the axle implementing or managing a security framework such as ISO or NIST. They are constantly surprised when their peers and CEO do not see them as partners in helping the business succeed.

To succeed as a security leader, you must develop relationships with your peers and the leaders of all the primary business functions (finance, HR, sales, production, etc.). If you have not already done so, schedule time with each functional leader to discuss what their group does and any pain points or concerns they may have regarding cybersecurity.

During these conversations, you should accomplish three things: discover what matters to the company's leaders, uncover what makes your organization a desirable target for attackers, and build the foundation of trust you will need to be an effective cybersecurity leader.

Here are the topics you should address, tailoring your questions to the role of the person you speak with. Don't approach this as a checklist to be reviewed in full with everyone you talk with, but tailor your questions as applicable to the role of the person you are speaking with.

2.1 Purpose

What is the organization's core reason for being—why does it exist? Knowing this will help you align your cybersecurity strategy with the organization's business strategy and communicate the value of cybersecurity to your business stakeholders. This is also an area where new security leaders often trip up, as they have an unstated assumption about the importance of cybersecurity that is frequently not shared by the company's leadership.

You may think the company's purpose is articulated in its mission statement and related marketing, but you could be wrong. For commercial companies, the answer is almost always making money first and foremost. However, even then, you can trip up if you don't understand your leaders' motivations and plans.

A typical example is in startup companies, where making money involves balancing acquiring customers, delivering products or services, and growing the company. It can also include plans to sell themselves to another company rather than continue the slog of multiple funding rounds. In these instances, the focus of the leadership is typically on "just enough" security to assuage the concerns of potential customers, not to develop the comprehensive security program that you may feel is necessary. Understanding this is key to surviving and growing with the company.

You can develop your understanding of your organization's purpose by using the following questions in your discussions (don't be surprised when you get divergent, if not conflicting, answers):

- What are the mission, vision, and values of the business?
- What are the goals and objectives of the business?
- What are the essential products and services that the business offers?

- Who are the customers and markets that the business serves?
- What are the business's competitive advantages and differentiations?
- What are the challenges and risks that the company faces?

2.2 Crown Jewels

What are your company's crown jewels? These are the systems and information that your business cannot survive without. They are the most critical assets that your business depends on, and threat actors will target them to steal, destroy, or exploit them.

To understand what these crown jewels are for your company, you can ask such questions such as:

- What are the key processes and functions that the organization performs?
- What systems and applications support these processes and functions?
- What data and information do these systems and applications generate, process, store, or transmit?
- How valuable, sensitive, or confidential is this data and information?
- How would the business be impacted if this data and information, or the systems that support them, were compromised, lost, or unavailable?

2.3 Revenues

How does your company make money, and from whom? Understanding your business's revenue streams and sources will help ensure your information security program is focused on protecting them.

To understand this, you should ask your business leaders questions like:

- What are the business's main revenue streams and sources?
- Who are the key customers and segments that our business serves?
- What are the expectations and requirements of these customers and segments?
- How does our business retain and grow these customers and segments?

2.4 Sensitive Information

What is the sensitive information, whether regulated information such as personally identifiable information (PII), patient data, financial records, or intellectual property, that the company stores, accesses, or processes? You will need this knowledge to understand your business's legal, regulatory, and contractual obligations and liabilities, how cybersecurity can ensure compliance with them, and, more importantly, the critical assets most attackers want to steal or destroy.

The broad categories of sensitive data you want to identify are:

- **Personal Identifiable Information (PII):** This includes information that can be used to identify an individual, such as name, date of birth, Social Security number, and driver's license number.
- **Protected Health Information (PHI):** This includes any information related to an individual's health or medical treatment, such as medical records, test results, and health insurance information.
- **Financial Information:** This includes any information related to an individual's finances, such as bank account numbers, credit card numbers, and tax identification numbers. This also includes information about the company's financials, especially for public companies, as this information can be used for insider trading and stock price manipulation.
- **Cardholder Data:** Cardholder data is any information related to an individual's payment card used to conduct a transaction, such as credit or debit card data. This can include the cardholder's name, card number, expiration date, and security code (CVV/CVC). In some cases, cardholder data may also include other personal information, such as the cardholder's address or phone number, if collected as part of the transaction process.
- **Intellectual Property (IP):** Any proprietary information or trade secrets an organization owns, such as formulas, designs, or processes.
- **Confidential Business Information:** This includes any information that is confidential or proprietary to an organization, such as business plans, customer lists, and marketing strategies.

- **National Security Information:** This includes any information classified or sensitive to national security interests, such as military secrets, intelligence information, and diplomatic communications.

Furthermore, don't be surprised if, in your conversations, you uncover regulatory or other obligations that the company's leaders were unaware of. Don't dig too deep into these in these conversations; note them as areas for further discussion and exploration. You don't want to be perceived from the start as Dr. Doom – you are trying to establish a positive relationship with all the company's leaders.

To understand what the organization considers sensitive information, you should ask your business leaders questions such as:

- What information does the company store, process, or manage that you consider especially sensitive?
- What are the sources and destinations for this information?
- What are our customers, partners or key stakeholders' concerns regarding our management of this sensitive information?
- Are there any laws, regulations, or contracts that apply to this sensitive information, such as CMMC, HIPAA, or PCI DSS?

In your conversations, consider the reporting requirements that your organization may face in the event of an incident involving this sensitive information. Depending on your industry, you may be subject to multiple reporting requirements (multi-state and federal) with different reporting triggers and deadlines. Again, don't be surprised if you get blank stares when you bring this up in your conversations - even if you don't get answers to your questions, you are planting seeds in executives' minds regarding the complexities that need to be addressed by cybersecurity.

2.5 Key Partners

What are the critical partner/supplier relationships and obligations? Cybersecurity threats impact more than just your company's internal IT environment. Not only do you need to ensure that you know where your company's data is being used or accessed by third parties, but you also need to know if you are doing the same for other companies and thus may have contractual security obligations to them.[2]

To find this out, you can ask:

- Who are the key partners/suppliers our business works with that are critical to our operations?
- What are the cybersecurity expectations and agreements we have with these partners/suppliers (don't be surprised if the people you ask do not know—that's okay, as you are planting a seed in their minds that this is something to be addressed)?
- Are we managing or processing sensitive information for another company, and if so, what are our security obligations to them (including the right to audit, notification of incidents, data destruction/return, etc.)?

2.6 Cybersecurity History

Finally, you must understand your organization's cybersecurity challenges, including past cyber breaches, current or anticipated regulatory or contractual cyber obligations, or cyber concerns from the board, key stakeholders, or customers. Without this knowledge, you are walking through a minefield of unknown expectations.

To investigate this area, ask questions like:

- Have we had cyber breaches or incidents that required reporting to the public, regulators, or partners?
- If we did, how did these breaches or incidents affect the organization in terms of financial, operational, reputational, or legal implications?
- What actions were taken in response to these incidents regarding new initiatives, funding, etc.?
- Have any cybersecurity concerns been expressed by the board, key customers, or regulators? If so, what were they? To date, how have they been addressed?
- How does the person you are speaking with perceive the current state of the cybersecurity program? Helpful (reduces risk, addresses outside concerns), harmful (a blocker to the business), or not something they even think about?
- How does your organization align and prioritize its cybersecurity strategy and initiatives in light of these concerns or expectations?
- How does your organization communicate and demonstrate its cybersecurity value and achievements to its stakeholders or customers?

THESE DISCUSSIONS SHOULD PROVIDE you with a solid "macro" level view of the organization and the assets critical to its survival and most likely to be targeted by your enemies. In addition, conducting these interviews will also build awareness of the need for information security in the minds of those you are interviewing, which will help you gain support for your information security plan, as we will review in the chapter "Present Your Plan".

3

UNDERSTAND THEIR THREATS

> The greatest danger lies in not recognizing the threat until it is too late.
>
> — LAO TZU

NOW THAT YOU know your enemies and what your organization has or does that makes it an attractive target, you next need to know the threats your enemy's attacks pose to your organization. These threats and their associated business impacts define the need for an information security program.

All too often, when discussing what could happen should a company suffer a cyber attack, security leaders use terms that sound like techno-babble when discussing the details with executives and board members. This is frustrating for business leaders who do not have a technical background, and it damages executive support for the security function. If the business leaders cannot understand "why" you need funding in terms that they understand, you are not likely to get it.

When engaging in risk discussions with your business leaders, it is crucial to use language that they can easily comprehend. This approach not only ensures that the conversation is inclusive but also respects their non-technical background. It effectively communicates your organization's threats, while making them feel understood and respected.

For instance, discussing your organization's potential vulnerability to malware introduced via malicious email attachments, which could lead to internal compromise and data exfiltration, might not resonate with most non-technical executives. However, presenting the fact that weak controls could make your company susceptible to ransomware attacks and backing it up with industry-specific statistics on the frequency of such attacks can significantly strengthen your argument and instill confidence in the need for a security program.

3.1 Extortion

Extortion is the threat of harm to a company or individual to extract money or some concession. Ransomware is the most well-known cyber-related extortion example occurring today.[1] In these attacks, companies are forced to pay a ransom to regain access to their data or not have it released to the public (called double extortion ransomware).[2] Another example of cyber extortion is denial-of-service attacks, where companies are extorted to pay a ransom to regain access to their online services.[3]

The latest evolution of ransomware tactics is so-called triple extortion attacks.[4] In addition to the attacks against data and the extortion to not release stolen information, companies are further attacked with a distributed denial-of-service attack or threatened with exposing the stolen information to their key customers, stakeholders or regulators.

According to the 2023 Verizon Data Breach Investigations Report, ransomware was involved in 24% of the breaches compiled in the report.[5] Government agencies,[6] businesses of all sizes, and even companies designated as critical infrastructure[7] have all fallen prey to ransomware. This universal threat underscores that no organization is immune to this attack, emphasizing the urgency and concern that all business executives and IT professionals should feel to protect against it.

Not only are ransomware attacks pervasive, but they can also be deadly. Businesses can be severely disrupted if not shuttered for good, with outages extended for weeks, if not months.[8] In addition, lives have been lost because of the disruption of IT systems caused by attacks on healthcare institutions.[9]

It's important to note that the typical expectation of most executives that their cyber insurance would cover the losses or damages from a ransomware attack is increasingly at risk. Insurance companies have expressly excluded ransomware from their policies in some instances,[10] and where insurance is still available, the premiums are skyrocketing.[11]

3.2 Fraud

Fraud is theft that uses manipulation and deceit rather than force, as in a robbery. Fraud is a cyber-related risk because criminals typically use phishing, social engineering, and other means to gain employee credentials or trick employees into providing access to sensitive information necessary to commit fraud.

These fraudulent activities can manifest in several forms,

each uniquely exploiting different vulnerabilities within business processes and human tendencies. The following are some key types of fraud that businesses must be vigilant against, though this is not an exhaustive list:

- **Business Email Compromise (BEC)** is a sophisticated scam targeting companies that conduct wire transfers, particularly those with international suppliers. This scam often involves cybercriminals hacking or spoofing email accounts of high-level executives or finance employees to initiate unauthorized transfers of funds. The FBI's 2023 Internet Crime Report indicated that reported BEC schemes resulted in over $2.9 billion in losses, a 20% increase over the previous two years.[12] An example of this type of attack is the 2018 €2 million loss for the Italian Series A football team, which made the final installment of a football player's transfer to the bank account of a scammer who used a malicious email to misdirect the payment.[13]
- **Supplier fraud** involves fraudulent activities perpetrated by entities posing as new or existing suppliers. These can include fake invoices, phishing attacks under the guise of vendor communications, or manipulation of supply chain systems to divert payments. A notable instance occurred in 2015 when Ubiquiti Networks reported a loss of $46.7 million due to fraudulent payment requests from purported suppliers.[14]
- **Disaster scams** spike following natural calamities or global crises when fraudsters exploit the chaotic environment and people's goodwill. These scams

can involve fake charity drives, phishing emails seeking donations, or selling non-existent emergency supplies. During the COVID-19 pandemic, for instance, there was a surge in fraudulent activities ranging from selling fake vaccines to phishing campaigns exploiting the crisis.[15]

- **CEO fraud** is a specific type of BEC in which attackers impersonate the CEO or another high-ranking executive to authorize fraudulent financial transactions. The fraudsters typically send an email from an executive's spoofed or hacked email account to a staff member with the authority to transfer funds requesting an urgent wire transfer to a fraudulent account.[16] The urgency and apparent high-level approval often bypass normal verification processes.
- **Payment diversion fraud** involves redirecting legitimate payments to fraudulent accounts. Attackers usually achieve this by compromising email accounts or communication channels to intercept and alter payment instructions. A UK retailer, for example, paid a fraudster £1.1 million, thinking they were paying a legitimate supplier.[17]
- **Payroll fraud** occurs when employees or external actors manipulate payroll systems to redirect payments or create fictitious employees to receive funds. This fraud can be particularly damaging, resulting in financial losses and undermining employee trust and morale.[18]
- **Invoice fraud** includes submitting fake invoices for payment, duplicating invoices, or inflating invoice

> amounts. It typically involves internal actors or collusion between employees and external parties. Google and Facebook reported an instance of invoice fraud in 2017 when they lost a combined $100 million to a Lithuanian scammer who forged invoice emails.[19]

Smaller organizations, in particular, are targets of fraud because they lack the controls or experience to spot sophisticated attacks.[20] In contrast, large organizations are targets because the size of their organizations combined with faulty internal controls makes it difficult to spot the attacks until it is too late.[21]

Exacerbating this problem is the increase in people working remotely. As more work is conducted via phone and email, workers have more difficulty verifying that the person they are dealing with is legitimate, not to mention the issues that exist when workers work from insecure networks, which is to say, most home network environments.[22]

Finally, advances in technology, in the form of AI-powered phishing emails, voice impersonations[23] and now video impersonations,[24] are making the attacks of scammers even more effective and more challenging to spot and are already leading to losses for companies and individuals who fall for these attacks.

3.3 Theft

This category of cyber threats involves unauthorized access, theft, or misuse of an organization's data or computing resources, typically for financial gain. The consequences of these actions can be severe, leading to significant economic losses and damage to the organization's reputation. Here are just a few examples of how data theft can occur:

- **Data breaches** are incidents where sensitive, protected, or confidential data is accessed or disclosed without authorization. These breaches can involve personal customer information, employee records, trade secrets, or intellectual property. The potential for personal harm is significant, as seen in the 2018 Marriott International data breach, which exposed the personal data of up to 500 million guests and is a stark example of the scale and impact these breaches can have on an organization.[25]
- **Intellectual property theft** involves the unauthorized access and theft of a company's intellectual property (IP), such as patents, trade secrets, and proprietary information. Intellectual property theft can significantly undermine a company's competitive advantage and long-term profitability. A notable case was a Chinese competitor's theft of wind turbine technology from American Superconductor, leading to a loss of more than $1 billion in shareholder value.[26]
- **Insider threats** are posed by individuals within the organization, such as employees or contractors, who have access to sensitive information and steal it for personal gain. It's crucial to remember that everyone within the organization has a role in preventing data theft, as seen when a former Google employee stole thousands of files that he used to start his own company, which Uber later bought.[27]

- **Resource hijacking** occurs when attackers gain control of a company's resources, such as servers or cloud-based services, and use them for their purposes. This often involves using the hijacked resources for cryptocurrency mining, also known as cryptojacking.[28] The Smominru botnet, for example, hijacked over half a million machines to mine Monero cryptocurrency, earning up to $3.6 million in illicit gains.[29]
- **Third-party data breaches** exploit an organization's vendors, who have access to an organization's data or infrastructure to facilitate theft. The Target data breach in 2013, where attackers gained access to Target's network through an HVAC contractor, is a prime example of data theft leveraging a vulnerable third party.[30]
- **Cloud storage breaches** are increasingly common as more businesses move to cloud solutions. Misconfigured cloud storage, weak access controls, and insufficient security practices can lead to significant data breaches. The Capital One breach in 2019, involving unauthorized access to over 100 million Capital One credit applications stored on a cloud server, is an example of the significant risks associated with storing data in the cloud.[31]
- **Mobile device data theft** is also increasingly common as mobile devices are more widely used within organizations. This can occur through lost or stolen devices, malware, or insecure apps. The theft of sensitive data from mobile devices can lead to significant security breaches and data losses, which

are increasingly occurring in healthcare organizations.[32]

3.4 Espionage

Cyber espionage is the illegal and unauthorized access to confidential information using electronic means, typically for political, economic, or military advantage. State-sponsored actors are usually involved in cyber espionage, using sophisticated tools and techniques to evade detection and attribution. China, Russia, North Korea, and Iran are the leading nations known to engage in cyber espionage against the U.S.[33]

Nation-state actors often employ criminal organizations as proxies in cyber espionage to maintain plausible deniability and avoid attribution.[34] Criminal organizations can carry out cyber espionage campaigns in exchange for financial gain, and they can use their expertise and networks to carry out attacks that are more difficult to detect and trace back to the nation-state actor.

One notable example is the North Korean government's use of the Lazarus Group.[35] The Lazarus Group has been linked to multiple high-profile cyberattacks, including the 2014 Sony Pictures hack and the 2016 Bangladesh Bank heist. While the Lazarus Group is believed to operate independently, it is widely considered to have ties to the North Korean government and to carry out cyber espionage campaigns on its behalf.

Another example is the Chinese government's use of the APT10 group. APT10 is a sophisticated hacking group linked to multiple cyber espionage campaigns against United States, European, and Asian organizations. It is believed to be part of a more extensive network of Chinese hacking groups that carry out cyber espionage campaigns on behalf of the Chinese government.

3.5 Sabotage

Cyber sabotage is a deliberate and malicious attack on an organization's infrastructure that causes economic or direct harm. It can impact businesses and critical infrastructure, such as power grids, transportation systems, communication networks, or financial institutions. Any threat actor group, from nation-states to cyberterrorists, can conduct these attacks, albeit with different motivations. Examples of cyber sabotage attacks include:

The 2012 Aramco cyberattack was a sophisticated and coordinated attack that targeted Saudi Aramco, one of the largest oil companies in the world.[36] The attackers used a virus called Shamoon to infect and wipe data from more than 30,000 computers belonging to the company by overwriting the master boot record of the infected computers, rendering them inoperable. The attack caused significant damage to the company's computer systems and resulted in the shutdown of its operations for several days. The attack was attributed to a state-sponsored hacking group from Iran.

Another example is the cyberattacks on critical infrastructure both Russia and Ukraine have been conducting against each other since the start of hostilities in 2022. One instance was the disruption of part of Ukraine's power grid in late 2022, attributed to Russian hackers.[37]

Finally, a massive and ongoing campaign against US critical infrastructure has recently been attributed to China state-sponsored hackers.[38] Targets span the communications, manufacturing, utility, transportation, construction, maritime, government, information technology, and education sectors.[39] As part of these attacks, hundreds of thousands of routers used by small businesses and individuals were compromised and used as part of the attack infrastructure.[40]

NOW THAT YOU know the threats you will likely face, it is time to pick those you feel are more likely to occur based on your company and industry. You will use this information and the discussion on likely threat actors/enemies to frame the conversation with your board and leadership on the need for and challenges of the information security program. Your strategic risks are the threats you have determined to be the highest risk (most likely and most impactful) to your organization.

Considering the ransomware epidemic occurring across all industries and government agencies, extortion will likely top the list for any company; the rest are more nuanced. An excellent way to develop these rankings is to hold meetings with the stakeholders involved, explain the options, provide examples, and then ask them to rank the options from most concerning to least concerning.

Don't worry about the exact ranking order. It is better to separate the threats into two groups: your primary concerns and your secondary concerns. Nowadays, no organization can be considered safe from any of these threats.

Your information security program should focus on preventing the first group of threats. As you evaluate controls, keep these at the top of your mind regarding the control's ability to mitigate these threats.

Once you have explained the threat actors and threats threatening the business, some executives will order you to ensure that all of these threats are countered. If this is the case, you need to remind them that while countering all the threats is your end goal, the order in which you do so should be guided by mitigating those that are most likely and impactful first. In other words, those that represent the highest risk to the organization.

Do not fall into the trap of attempting to "boil the ocean." That is, address every issue, everywhere, all at once. This will always fail, guaranteeing that if anything is done, it will be the easiest task rather than the most important in mitigating risk.

Now that you have determined your most likely enemies and the threats they pose to your organization, it's time to develop your plan to counter these threats. This is the objective of Part Two, Tactical Risk Management, which I will cover next.

PART II

TACTICAL RISK MANAGEMENT

GET PREPARED

Thus, what is of supreme importance in war is to attack the enemy's strategy.

— SUN TZU, THE ART OF WAR

4

KNOW THEIR METHODS

> Never attempt to win by force what can be won by deception.
>
> — NICCOLÒ MACHIAVELLI, THE PRINCE

UNDERSTANDING the most likely threats to your business is crucial, but grasping the common methods these attacks employ is equally essential. In this chapter, I will delve into the primary means by which all threat actors target organizations. This knowledge will provide a solid foundation for understanding the controls necessary to counter these threats.

All cyber-attacks fall into one of two broad categories: compromise the person and compromise the system. Attackers will use multiple means of attack and chain these together to meet their objectives. Each attack method should be viewed as a link in a chain of events that narrates the story of an attack.[1]

The sequence in which they appear in the chain depends on the attacker and their aim.

4.1 Compromise the Person

People, either employees or contractors, working for you or at third parties whose products or services you use, can be compromised to perform or facilitate cyberattacks. These attacks occur in two primary ways, which are:

Social Engineering

Social engineering is the term for the deception or manipulation of individuals to get them to do something that facilitates or enables a fraudulent act. This deception can occur via an email or fake website (phishing),[2] phone call (vishing),[3] or text message (smishing).[4] All these methods are intended to trick an employee into performing some action supporting the attacker's objectives.

This can range from enticing the user into executing malware disguised as an innocuous attachment in an email,[5] to clicking on a link to direct them to a counterfeit website where their login credentials are stolen,[6] to tricking them into providing sensitive information or otherwise performing some action to help the attacker commit fraud. [7]

It is essential to realize that it is not just your employees who are at risk of social engineering; it is also the employees of any organization who have access to your environment or host your data.[8] Compromising a contractor's laptop with physical access to your environment is just as valuable for an attacker as compromising one of your employees, and it is often easier if the firm that the contractor comes from does not have as robust cyber controls as your organization.

Physical Attacks

From a cyber perspective, physical attacks describe instances where the attacker physically interacts with you or the business location to conduct their attack. This can range from tailgating[9] (following an authorized user into an area without clearance) to access a secure area, to using social engineering (pretending to be a customer or repair person) so that they can enter the company offices and install a malicious device[10] that allows attackers remote access to the company's internal network.

USB dropping is another attack that mixes social engineering (tricking the individual) and physical (leaving malicious USB sticks where they are likely to be picked up and hopefully inserted into company computers).[11] Once a malicious USB is plugged into a company computer system, it can do anything from installing malware to steal information to physically damaging or destroying the computer it is plugged into by delivering an electrical surge, a so-called "USB Killer".[12]

4.2 Compromise the System

All electronic systems, whether you own them or use those a third party provides, can be compromised to perform or facilitate cyberattacks against your business. The most common means by which these attacks occur are:

Execute Malware

Malware is any software intended to steal information or manipulate systems for malicious purposes (ransomware is an example of malicious software).[13] While malware is often introduced into systems by tricking users into executing it via social engineering (malicious email attachments, for example), it can also be embedded into products for supply chain attacks by compromising the open-source software used in these products.[14] Attackers can then use the access that this malicious code provides to penetrate the systems and networks of the customers using the compromised product. The SolarWinds attack is a perfect example of nation-state attackers compromising a commercial product to attack the users of the product.[15]

Exploit Misconfigurations

Information technology systems and networks are highly complex and are getting more so daily, especially with the move to cloud computing, which adds another dimension of complexity. For every software, hardware, and networking component, configuration is usually required to set it up to perform its intended function. The issue arises when this configuration is not done, and the component operates with default (and typically insecure) settings, or it is done by someone who does not know what they are doing.[16] The result is that hackers often have an open door to steal data or compromise systems by taking advantage of these misconfigurations.[17]

Exploit Vulnerabilities

Vulnerabilities differ from misconfigurations in that they are flaws in the component (hardware or software) that leave it open to misuse. In contrast, misconfigurations are not flaws but mistakes in setting up the system properly. Scanning for and exploiting misconfigurations and vulnerabilities is a primary method hackers use to compromise systems, simply because these issues are so prevalent.

It is not unusual for a large corporation with tens of thousands of computer and network devices to have millions of known and unknown vulnerabilities, ranging from minor to critical (if the hacker exploits the vulnerability, they will have complete control of the device). Making this problem even worse is that many of these vulnerabilities, once discovered, have no patch available.[18]

This can be because the manufacturer has either gone out of business, does not have the resources or interest to resolve the issue, or considers the product end-of-life.[19] As a result, companies with these obsolete systems have the equivalent of ticking time bombs in their environment, just waiting for a malicious hacker to take advantage of them.

Worse is when the hackers themselves discover the vulnerability before the manufacturer, resulting in what is known as a Zero-Day vulnerability.[20] When these are announced, a mad rush starts as companies try to take whatever remedial measures may be available (if there are any) as the attackers swarm vulnerable systems to exploit the open hole. Nation-states, in particular, collect zero-day vulnerabilities for their use in foreign and domestic surveillance, espionage, and sabotage.[21]

A last point to understand is that a company can have a perfect vulnerability management system, scanning and patching vulnerabilities daily (a practical impossibility for any commercial organization), and still be exposed. It is estimated that over 100,000 vulnerabilities are undiscovered, which means that no amount of effort can provide perfect protection (which is also an impossibility because of zero-day vulnerabilities).[22] The most you can do is have a good process and strive to improve it constantly. You are on a battlefield, and bullets are flying. You can do your best to protect yourself, but sometimes, your best will just not be good enough.

Exploit Design Flaws

Hacking is a generic term for all attempts to compromise computer systems or networks to exploit or steal information from them. While it can describe activities as basic as running scanners to detect vulnerabilities and misconfigurations by unsophisticated criminals (script kiddies), true hackers will use all the avenues described above, combined with their ability to write custom code or develop custom hardware, to take advantage of previously undetected design flaws and subsequently penetrate the systems they are targeting.

This is where zero-day vulnerabilities are discovered and used by attackers to penetrate systems through holes that the system owners did not even know existed. Nation-state hackers excel in this area, as they find and use these vulnerabilities for espionage and sabotage, never letting the public know they even exist.[23]

NOW THAT YOU understand the methods attackers use to carry out their threats, it's time to examine the controls you will need to implement in your organization to counter these methods. The next chapter, "Assess Your Defenses", will focus on this.

5

ASSESS YOUR DEFENSES

> The art of war teaches us to rely not on the likelihood of the enemy's not coming, but on our own readiness to receive him; not on the chance of his not attacking, but rather on the fact that we have made our position unassailable.
>
> — SUN TZU, THE ART OF WAR

So you have identified your likely attackers, the threats that their attacks pose, and how, at a high level, they will carry out these attacks. Now what? Now, we come to the meat of your information security program, which is the countermeasures (controls) you need to implement to defend against these threats.

To the degree that you have gaps in either the implementation or maturity of these controls, these represent your tactical risks. You will manage and track these risks by defining and implementing your information security plan.

WHAT ARE CONTROLS? Controls are the measures you put in place to defend against the methods attackers use to exploit the vulnerabilities you may have in your environment. For example, if you do not have a policy requirement that your users use complex passwords or you do not have the technology to enforce such complexity, these are vulnerabilities. Something that attackers can exploit.

The controls you implement to counter this vulnerability can include anything from a policy mandating that users create and use complex passwords, to technology (such as your Active Directory settings) to enforce this policy requirement. When implementing controls, your intent should be to mitigate the threats by eliminating the vulnerability (if possible) or at least making it more difficult for an attacker to exploit it.

To determine the controls required to counter cybersecurity threats, there are numerous frameworks for cybersecurity that I have already mentioned in prior chapters. Choosing the right framework for your security program is a crucial decision and the stage where, in my experience, many information security programs typically go off the rails. Rather than focusing on the fundamental controls, I have often seen security leaders or CEOs mandate the most comprehensive set of information security controls possible. The thinking is that if every possibility for failure is eliminated (by mandate), failure is impossible (or at least highly unlikely). Instead, this approach is a prescription for failure under the metaphor of "Boiling the ocean".

When you attempt to cover every possible control, you will likely spend your time and effort focused on mandated activities that are easy, cheap, or sexy (in a technology sense) to show progress instead of focusing on activities with the greatest risk mitigation benefit. Implementing these controls, such as asset management and patching, is often difficult, expensive (from a manpower perspective), and not sexy.

When reviewing the history of major data breaches in large corporations, the root cause can often be traced back to a failure to implement and manage fundamental security controls properly.[1] Examples of these fundamental control mistakes include ineffective patching, failure to configure equipment properly, or failure to change default credentials.[2]

What is curious is that these incidents occur in organizations that spend tens if not hundreds of millions of dollars on information security and have security organizations with hundreds if not thousands of dedicated security staff. How can these events happen so frequently with all this attention to security?

They happen because people mistake increasing complexity for increasing security. Just as being compliant does not mean you are secure, being busy (implementing and managing controls) does not mean your efforts are effective (in reducing risk).

Governments and regulators perpetuate this problem by adding more security regulations and requirements after every significant security breach. Because of the increased distraction these mandates bring, security decreases rather than increases within impacted organizations.

So what can you do? Before doing anything else, start with the fundamentals and ensure you are managing them well!

5.1 Build A Strong Foundation

So, what are the fundamental security controls? You could parse through the various cybersecurity frameworks to identify the common controls, match them to the exploited vulnerabilities they mitigate as determined by the recent history of breaches, and then decide how to ensure that you have implemented the controls you selected effectively. Or you can leverage the work of the Center for Internet Security, Inc. (CIS), a community-driven nonprofit responsible for the CIS Controls® and CIS Benchmarks™, globally recognized best practices for securing IT systems and data. CIS leads a global community of IT professionals to continuously evolve these standards and provide products and services to protect against emerging and evolving threats.

The CIS Controls, now in Version 8, have been mapped against the MITRE Enterprise ATT&CK® framework v8.2,[3] the industry-accepted way to describe the individual technical details of a cyber-attack, which, in turn, can be used to structure and test controls to mitigate these attacks. This mapping, documented in the CIS Community Defense Model (CDM) V2,[4] showed that the CIS Controls effectively defended against 86% of the techniques found in the ATT&CK framework. In particular, the CIS Controls were highly effective in protecting against the top five attack types found in the industry threat and breach data: Malware, Ransomware, Web Application Hacking, Insider and Privilege Misuse, and Targeted Intrusions.

The CIS Controls themselves are (as of this writing) 18 top-level controls addressing the most critical requirements for effective defense. These controls are, in turn, subdivided into 153 sub-controls, which the CIS refers to as Safeguards. These Safeguards are categorized into three implementation groups,

IG1 through IG3, with IG1 consisting of 56 Safeguards that CIS considers the foundational set of security measures that every enterprise should have to defend against the most common attacks.[5]

The IG2 group consists of 74 additional Safeguards intended for security functions in organizations with regulatory requirements concerning data security and tight operational deadlines, allowing only limited downtime.[6]

Finally, the IG3 group added 23 additional Safeguards, building on those in IG1 and IG2. These controls are for the most sensitive organizations with stringent availability, confidentiality, and integrity requirements and where successful attacks can cause significant public harm.[7]

The CIS Controls are increasingly used as a standard of good practice for all organizations. Government agencies commonly use them to audit themselves to ensure adequate cybersecurity measures are in place.[8]

But what if, you say, you are already compliant with an established industry framework such as PCI, FedRamp, CMMC, or any of the other security control frameworks in existence? I can tell you from experience that it is a very rare organization that fully complies with the standards it has been certified to, regardless of its size or maturity.

I have observed that as organizations grow their security departments and seek alignment with more complex security control frameworks, their focus often shifts from implementing and managing controls to ensuring that they can demonstrate compliance with the standards they have adopted or been mandated to follow. This is a practical reality in large organizations and has much to do with the hamster wheel I discussed in the introduction. Security organizations, especially large ones, are particularly challenged to keep up with the management of existing controls and the implementation of new ones. Their focus in audits is often on demonstrating that they are at least meeting the spirit of the intended control, if not the letter of the requirements.

This means that regardless of the maturity of your security organization, you should baseline your program against the CIS Controls and focus your efforts on closing the gaps you identify there, before addressing any other gaps you may have relative to formal compliance with a "required" control framework. The result will be an information security program that evolves into an effective and compliant one instead of a compliant but ineffective security program (which far too many are, as demonstrated by the continuous rise in security incidents and data breaches).

5.2 Measure Your Maturity

So, how do you determine if you are meeting these requirements? CIS again provides a wealth of tools and information. You can assess your organization's maturity against the CIS Controls by using the self-assessment tool CIS provides, the CIS Controls Self Assessment Tool (CIS CSAT).[9] This tool helps organizations assess, track, and prioritize their implementation of the CIS controls.

CIS CSAT has two versions: a CIS-hosted version and an on-premises version for CIS SecureSuite members called CIS CSAT Pro. The CSAT Pro version also provides additional capabilities for assessing and tracking progress for larger organizations, as well as distributing the workload of gathering assessment data.

Organizations begin the assessment process by defining the scope, including which parts of the organization will be assessed. Teams are set up within the CSAT platform with assigned roles and responsibilities if using the Pro version. During the execution phase, participants answer questions about each of the CIS Controls and Safeguards based on the current state of their cybersecurity practices, providing evidence to support their responses where possible.

As you assess your organization against the CIS Safeguard requirements, you will determine whether you have a Policy Defined, a Control Implemented, a Control Automated, and a Control Reported for each Safeguard. Your answer to each question determines the maturity of the Safeguard in your environment and the gaps you need to close relative to the specific Safeguard.

The CIS CSAT output includes a numerical implementation score reflecting the degree to which each control and Safeguard is implemented, detailed gap analysis reports, and dashboard visualizations representing the organization's cybersecurity posture. Additionally, the tool provides recommendations for improving Safeguard implementation where gaps exist and maintains documentation of the evidence supporting the assessment.

In addition, you can anonymously compare your results to industry averages, an essential component of your security plan proposal to your leadership. This is vitally important because, in my experience, one of the most often discussed aspects of your security budget proposal is the question, "What are our competitors doing?". Boards frequently use this heuristic to gauge whether your request aligns with industry expectations, as they may lack the expertise to review your proposal on its merits.

My intent in this book is not to provide details on the CIS Controls and Safeguards or the CSAT tool and assessment process. Excellent information on these topics is already available on the CIS website. Instead, I want you to understand that the CIS Controls are the most effective foundation (in terms of cyber threat risk reduction) on which to build your information security program. The CIS assessment process, which considers implementation from multiple perspectives, is far more robust than the typical control assessments conducted by most auditors.

ONCE YOU HAVE COMPLETED your CIS assessment, either manually or using the CIS CSAT, you should now know the gaps in your existing security program based on your maturity relative to the CIS Controls Safeguards.

Your next task is to prioritize these gaps. As previously discussed, the Implementation Groups (IG) within the CIS Controls signify a progressive maturity level of security tailored to the organization's needs. If your organization's regulatory and data requirements necessitate implementing the complete set of 153 Safeguards, then remediation of deficiencies should be prioritized based on these IG levels. In other words, start by addressing all IG1 gaps, followed by IG2 gaps, and so on.

It's important to understand that fully implementing CIS controls may be a multiyear journey, depending on your current maturity and available resources. Likewise, fully implementing all the controls may not be desirable depending on the risks to your organization. If you are a nonprofit with limited funding, then the IG1 Safeguards may be all you can afford to implement and maintain. Likewise, if you are a healthcare organization looking to obtain HiTrust certification, you will likely want to implement all the CIS Safeguards as part of your journey to eventual accreditation.

Prioritizing your needs and having solid reasoning for that prioritization will play a significant part in your budgetary request to management for your program's funding, which we will discuss in the chapter "Present Your Plan." People are a vital part of the plan you will be developing, as without staff, you will be unable to implement, configure, monitor and maintain the controls you have deemed necessary. To help you better understand your needs in this area, the next chapter discusses the minimum personnel requirements for a security organization so that you can factor these into your budget request.

6

ASSEMBLE YOUR TEAM

> All of the real heroes are not storybook combat fighters either. Every single man in this Army plays a vital role. Don't ever let up. Don't ever think that your job is unimportant. Every man has a job to do and he must do it. Every man is a vital link in the great chain.
>
> — GENERAL GEORGE S. PATTON

CLEARLY DEFINED responsibilities and scope of authority for security organizations vary widely across different companies and industries. To help address this challenge, I have developed an organizational framework for the information security function based on my experiences as CISO across multiple organizations. This should be considered a starting point for adapting to a particular environment as a model, not a set-in-stone guide for the division of responsibilities. It should not be deemed all-encompassing to define every possible function to be performed.

It is essential that you consider your organization's staffing needs at this point, as you will need to make this part of the information security plan you propose, as we will discuss in the next chapter, "Present Your Plan." In most cases, the gaps you identified in your risk assessment from the prior chapter will require people to create, configure, deploy, and manage the initiatives to close them.

From a budgetary perspective, people are typically the largest component of a security organization's budget, whether they are employees or you have outsourced a function to a third party, such as managed detection and response. Tying your staffing needs to your initiatives will ensure that your requests will be much more likely to be approved instead of just submitting an organization chart and requesting staff in isolation from the risk drivers for the ask.

The acronym DIOR stands for what I consider the critical responsibilities of any security organization: Define (the risk & plan), Implement (the plan), Operate (the plan), and Respond (to incidents). How these components are organized, as departments within the overall security organization, spread across multiple departments in a matrixed fashion, or outsourced to service providers, is highly dependent on where Information Security sits within the organization and the scope of authority of the Chief Information Security Officer (CISO).

6.1 Define

The overarching goal of any security organization should be to defend the company's brand and operations from cyber threats by implementing a risk-based cybersecurity program and plan, which is what this book is all about. The "Define" team, usually led by the CISO or the designated head of security, focuses on creating and maintaining this program. The "Office of the CISO" is also used to describe those on the Define team.

To accomplish the objective of "Define", the following functions should be considered:

- Understand the mission and objectives of the organization, as well as its operating constraints and regulatory requirements
- Identify the organization's critical assets, including systems and data. This is typically the role of IT; however, the Define team needs to ensure it is done correctly
- Determine the likely threats to the organization and its assets
- Create the policies and controls necessary to address these threats
- Monitor the effectiveness and adherence to the controls in reducing risk and update risk plans accordingly
- Train users on relevant cyber risks, maintain ongoing awareness, and test awareness to ensure the training is effective
- Govern the program by developing and maintaining relationships with key stakeholders, gaining consensus and support on the program and plans, and reporting progress

6.2 Implement

The Implement function implements your tactical security plan's security controls, processes, and technologies. All too often, controls are introduced without considering their impact on business/IT operations, worker productivity, or ongoing support costs. The Implement team is responsible for making sure this does not happen.

To accomplish the objective of "Implement", the following functions should be considered:

- Evaluate/recommend control technologies and processes needed to meet policy requirements and integrate them into the business environment
- Keep abreast of changes in technologies, processes, and the business that impact organizational security
- Review/approve changes in the organization's technology environment
- Develop secure architectures for technology deployment that are aligned with the organization's business needs
- Develop standards for the secure configuration of IT/IS products and services and update them annually and as needed
- Collaborate with IT operations to implement and configure security-related technologies

6.3 Operate

Effectively operating security technologies is a requirement all too often given short shrift in many organizations, and as a result, expected gains in risk reduction are rarely achieved.

Centralizing security operations and considering this cost in technology/process decisions ensures that costs/benefits are realistically addressed when evaluating alternatives.

To accomplish the objective of "Operate", the following functions should be considered (this is much larger than it looks, both in terms of staff and complexity, due to the number of technologies typically deployed):

- Operate and monitor the security suite of tools to prevent, detect, and respond to intrusions, attacks, and vulnerabilities
- Investigate events and raise confirmed or suspected intrusions with the Respond team as appropriate to the risk
- Coordinate with IT operations (internal or outsourced) to track and remediate identified issues and vulnerabilities

6.4 Respond

A security program that has not prepared for failure is a failed security program. Incidents are inevitable in any security organization, regardless of its size or budget. The degree to which organizations are ready for these events determines whether the organization survives, suffers significant costs, or even fails.

A common flaw in security organizations is that staffing response/recovery is not a dedicated function. This means that preparation and response are ad hoc at best, as they are left to the rest of the organization.

To accomplish the objective of "Respond", the following functions should be considered:

- Investigate possible attacks as raised by the Operate team and coordinate response, remediation, and recovery as appropriate
- Prepare for likely attack scenarios and train participants on what to expect and how to respond
- Monitor operational threat intelligence sources and alert the Define team as needed to significant new threats
- Test the organization's resiliency and effectiveness of controls periodically through internal and external testing activities
- Prepare and coordinate outside resources as needed to support response plans and activities, as well as being available for rapid response should a large-scale incident occur

WHEN ORGANIZING and staffing your security function, it's crucial to consider the unique personalities and interests of the people you hire. Security professionals, like those in the medical profession, often lean towards specific subspecialties that align with their personalities and interests. Attempting to mold every security professional into a generalist will lead to mediocrity and failure. Similarly, an information security organization without clear responsibilities and dedicated staff to fulfill them will not be effective in managing risk.

As you create your model organizational plan, list the responsibilities of each person or team in relation to the DIOR model. Ensure all groups are adequately staffed to perform the functions, particularly the operation teams. It is due to the size and cost of these teams that the function is usually outsourced to a service provider; however, know that you can never totally outsource this function — there needs to be someone(s) on your end to pick up the phone and conduct investigations when the service provider calls with a suspicious event.

In the next chapter, we will combine the tactical risks you have identified (your required initiatives) and staffing requirements to develop the plan you will present to your management team for approval and funding.

7

PRESENT YOUR PLAN

Plans are worthless, but planning is everything.

— GENERAL DWIGHT D. EISENHOWER

IF PLANS ARE WORTHLESS, why bother to plan? While you may not know how things will unfold, you can certainly anticipate what is most likely to happen and make the appropriate preparations and plans. Planning for those contingencies is the foundation for ensuring your organization's survival in the event of an attack. This is what Eisenhower was referring to in his famous quote.

The plan you create here is your security organization's operating plan. Unlike the information security program, which sets the purpose for the security function in terms of what it is protecting the organization from (the "why", in other words of the security function), the information security plan is a dynamic document that specifies the "how" of the initiatives you feel are necessary to protect the organization. These initiatives, driven by the gaps you have uncovered in key security controls within your organization, represent the tactical risks to your organization.

I will present this process assuming you will present the first information security plan for approval to the board or senior management. If you are not the first security leader and there is an existing approved plan and process in place, you will need to review what has been presented and approved in the past and modify it to meet what you see as the organization's needs.

You never want to surprise your board or management. Before presenting the deck, discuss any significant changes in format or content that the board is used to seeing with key board members or senior executives. Let me repeat: NO SURPRISES!

The remainder of this chapter assumes that you are making the first presentation to your board on the need for and requirements to implement an information security program. However, depending on your program's maturity and your organization's size, you must adjust these to your specific needs.

7.1 Board Reporting

While there are various opinions about what boards (or senior management, depending on your organization) need and want to hear regarding information security, let me distill it down to the minimum requirements based on my experience. This should not be considered the final word for developing your board presentation but rather the starting point that you should tailor based on what your board or senior management is used to or is requesting to see.

The Gramm-Leach-Bliley Act (GLBA), which became law in 1999, established, I believe, the foundational requirements for all information security programs.[1] A component of this legislation called the Safeguards Rule, specified minimum safeguards for the protection of consumer information and took effect in 2003, but after public comment, the FTC amended it in 2021 to make sure the Rule (now referred to as the "Final Rule") keeps pace with current technology. The Final Rule now specifies nine specific elements that an organization's information security program must include:[2]

- Designate a Qualified Individual to implement and supervise your company's information security program.
- Conduct a risk assessment.
- Design and implement safeguards to control the risks identified through your risk assessment.
- Regularly monitor and test the effectiveness of your safeguards.
- Train your staff.
- Monitor your service providers.
- Keep your information security program current.
- Create a written incident response plan.
- Require your Qualified Individual to report to your Board of Directors.

As the first requirement, the role of the Qualified Individual is pivotal in implementing and supervising your company's information security program. While this may be you, the reader, the other elements are all covered directly or indirectly in the other chapters of this book.

What does this have to do with your initial ask to management or the board and your ongoing reports to these groups? Everything. To quote the Final Rule regulations, 16 CFR 314.4(i):[3]

> Require your Qualified Individual to report in writing, regularly and at least annually, to your board of directors or equivalent governing body. If no such board of directors or equivalent governing body exists, such report shall be timely presented to a senior officer responsible for your information security program. The report shall include the following information:
>
> (1) The overall status of the information security program and your compliance with this part; and
>
> (2) Material matters related to the information security program, addressing issues such as risk assessment, risk management and control decisions, service provider arrangements, results of testing, security events or violations and management's responses thereto, and recommendations for changes in the information security program.

These requirements are what I recommend that you cover in your first and ongoing reports to either your board of directors or equivalent governing body as the regulation refers to them (usually the CEO and other c-suite executives). You will need to tailor this depending on whether this body has received prior reports regarding the information security program or if this is the first.

If this is the first, it is your opportunity to set the tone and format for these meetings to some extent. Before your formal meeting, you must always confirm your presentation content and format with someone from the board. Find a relevant member of the board, usually the chair of the audit or risk committee, to review what you intend to present and why. The fact that your presentation agenda aligns with federal regulations makes your job easier, as it assures board members that they are getting the information that regulators expect they should be receiving.

Also, if this is the first presentation the board has received on this topic, you may not have information yet on certain items, such as service provider arrangements, testing results, or security events. In that case, add placeholder slides to your deck discussing what you will be presenting on these topics in the future and what you intend to do in the interim to get that information (such as scheduling testing, stratifying service providers, reviewing contracts, etc.).

You must also confirm with the board members how much they want to see on these topics. This is another area where inexperienced CISOs often shoot themselves in the foot. The higher you go in an organization, the rule is "Less is More".

You should limit technical details in your main slides as much as possible unless you have a highly unusual board made up of technical and security experts. Even then, I recommend summarizing the information in the main body of the presentation and including any requested technical details in attachments or appendices.

7.2 Funding Request

The final component of your presentation addresses the last of the nine required elements: "recommendations for changes in the information security program." This is where you pitch for the initiatives needed to plug the gaps you identified in your CIS risk assessment.

A point often lost on CISOs is that they are working for a business, and one of the constraints of any business is that they only have so much time, resources and attention that they can devote to any particular topic. Regardless of how earth-shatteringly important a CISO may feel it is to close the gaps that they have identified, to the board and CEO, these are just another set of problems that must be prioritized against all the other urgent issues on their plate, such as acquiring and retaining customers at a reasonable profit margin.

This means that CISOs must leverage the risk prioritization of the initiatives they developed as part of their risk assessment and realize that it is the rare CISO who gets all they ask for. During the discussion of the requested changes to the information security program (i.e., your budgetary proposal), you need to discuss what gets done and what is deferred in terms of risks being mitigated and risks being accepted (for the moment, at least). The point is that just because the board does not approve funding risk remediation, it does not mean the risks do not exist just because they are ignoring them.

This point is critical because should the proverbial fecal matter hit the oscillating air mover due to one of the gaps you identified being exploited, you want to be able to show that you identified it to the board and that it was THEIR decision not to fund its remediation at this time. This is why one of the outputs from the board meeting needs to be a review of any agreements made in the meeting, including where the line was drawn regarding initiatives funded and initiatives deferred (or risks accepted).

When presenting your proposal for changes to the program, I recommend that you group what you are asking for into three buckets: essential security initiatives (Critical holes in the program — IG1 gaps — address immediately), urgent security initiatives (IG2 gaps high priority — schedule next), and recommended security initiatives (IG2 gaps lower priority — schedule later). These groupings should align with the Risk Appetite statement that you have either negotiated with the board or been told it is already defined (see the chapter "Manage Your Risks" for a discussion of Risk Appetite).

You want the board to feel that they have some input into the process and that you have thought through the issues before it gets to them, rather than presenting them with a laundry list of problems without clear prioritization. Doing so eliminates or reduces the discussion on the individual items and instead shifts the focus to macro-level risk mitigation. A critical point to remember is that, at the end of the day, what gets funded **is not your decision but belongs to the board and management**. You are simply advising what you think SHOULD be done and why.

The information you present to the board on these initiatives needs to include four key components: the operating cost (Opex) for the initiative, the capital cost (Capex), any required FTEs and their costs (which I usually break out from Opex), and finally an estimate of timing. NEVER give exact dates for initiatives, especially when you are new to the organization, unless you have psychic powers that allow you to predict the future accurately (in which case, what are you doing as a CISO?). Instead, provide a range (6-9 months, etc.), which you can refine in subsequent updates as your detailed project plans for each initiative are developed.

7.3 Policies Approval

Another item you need to get approval for that is not explicitly mentioned in the GLBA board report guidelines is the organization's information security policies. If you are new to the organization, you will likely develop them. If not, you may need to revise what is already in place.

In my experience, overly prescriptive information security policies are nothing more than spiked clubs that you hand to auditors and regulators to beat you with. Never create a detailed policy for a practice that you cannot meet! Instead, your information security policies should be prescriptive in terms of what has to be done yet flexible enough that you can grow into them by adjusting the associated standards while leaving the policy statements alone.

For example, vulnerability management. I have seen entire encyclopedias that are shorter than some vulnerability management policies. The result is a document nobody reads, nobody maintains, and which, for all practical purposes, is useless other than an artifact to provide to an auditor (so that they can use it as a club to beat you with).

A better alternative is a simple policy statement: "All information systems and devices shall be scanned for vulnerabilities regularly and patched for these vulnerabilities on a schedule commensurate with the risk posed by the vulnerability". You should also indicate what group(s) are responsible for carrying out the policy, but otherwise, that's all you need to say!

What is regular scanning? Or patching schedules based on risk? Define that as a separate standard, referencing your policy statement, and ensure that it is something you can meet. First, it could be scanning once a quarter and patching critical vulnerabilities within 30 days and all others within 90 days.

While this is undoubtedly weak, it is better than nothing, especially if you follow the standard you set for yourself. My point is, don't set yourself up to fail from the start. Adopt a standard, make it work, and THEN improve it. Don't set the standard that you will achieve world-class patching performance as your baseline. You are only setting yourself up for failure and audit findings.

If you don't establish realistic standards, those standards will come back to haunt you, not only in an audit but worse in legal proceedings when your stakeholders sue you for a breach caused by an unpatched system. You can be sure that the plaintiff's attorney will read your policies and standards and will be more than happy to point out that you were not following them, thus leading to the breach. The question then becomes: were you incompetent, clueless, or just lying? It's not a pretty position to be in.

If you develop your policies as I recommend for the controls you are implementing as defined by CIS, the result may be able to fit on just a few pages. This will result in a set of policies that will guide your program, not constrain it, and will be of such a length that everyone can actually read and understand them. Leave the details for the standards you develop to flesh out the "how" of meeting the policy objectives.

7.4 Kobayashi Maru

One final point regarding presenting your plan. For those unfamiliar with the Star Trek movies, the Kobayashi Maru references the film Star Trek II: The Wrath of Khan.[4] In the movie, it was presented as a no-win training exercise designed to test the judgment and character of Starfleet Academy cadets. As a CISO, you must recognize when you are in a Kobayashi Maru scenario.

CISOs have some of the highest rates of job burnout in the industry.[5] A big reason for this is not only the complexity and pressure of the role but also the feeling of helplessness that can come when you are not only not supported but actively fighting with your board and management to get recognition of the dangers facing the organization and the support you need to address them.

As the security leader, signs that you are in such a situation are when your plans are rejected outright or barely funded, management is blocking your access to the board, or your reports to the board are heavily redacted. While there may be valid reasons for these actions, as the designated person responsible for securing the organization's information, you need to decide whether the risk to you is worth staying for. You don't want to be the designated scapegoat for the company should a breach occur.[6]

With the SEC now personally pursuing CISOs when there is a security breach in public companies, as occurred to the CISO for SolarWinds,[7] CISOs, especially those in publicly listed companies, must take a hard look at their exposure. Even with company-provided liability insurance and a well-documented risk plan and risk register, you still need to contend with the possibility that your life may be turned upside down for years should a significant breach occur at your company.

Even for non-public companies, CISOs are often thrown under the bus when a breach occurs as a cover for management's failures to follow the CISO's recommendations. While this is not always true — being a CISO is challenging, and not everyone in the role is up to the task — I have seen it happen more often than not.

Does this mean you should cut and run at the first sign of pushback? Absolutely not! Instead, I am warning CISOs that they must be aware that sometimes they can be in an unwinnable position. When you realize this may be your situation, do you risk going down with the ship or head for the lifeboats? The choice and decision are yours.

Next, in Part Three, "Operational Risk Management," we will discuss how to use risk to manage your information security plan effectively.

PART III

OPERATIONAL RISK MANAGEMENT

DEFEND YOURSELF

No plan of operations extends with any certainty beyond the first encounter with the main enemy forces.

— PRUSSIAN FIELD MARSHAL HELMUTH VON MOLTKE

8

EXECUTE YOUR PLAN

> A good plan violently executed now is better than a perfect plan executed next week.
>
> — GENERAL GEORGE S. PATTON

NOW THAT YOUR information security program and plan have been defined and approved, it's time to implement them. At this stage, security leaders must use all their skills to juggle multiple balls while riding a unicycle. You will be dodging incoming bullets from threat actors attacking your organization while dealing with the internal politics of implementing your plan.

This is where operational risk management comes into play. You will use your risk register to track and manage the risks you discover in the day-to-day management of your plan. These risks will typically fall into one of three categories: self-identified variances to established policies and standards, variances discovered by third parties, or variances discovered as a result of incidents. I'll go over each of these in the following chapters.

However, before we explore those topics in further detail, I first want to cover the key aspects of implementing the information security plan you received approval for in Part Two.

8.1 People

Before you can effectively manage an information security function, you must first assemble a team. While I discussed the essential functions you needed in your organization in the chapter "Assemble Your Team" so that you could incorporate them into your budget, now you need to hire and assemble that team based on the budget you got approved. The team you build will likely consist of a mix of people you hire and those you contract to perform specific functions from managed service providers.

Given the difficulty of finding and keeping talented security professionals, it makes sense to outsource those functions that are most costly to manage and retain to third parties, such as the bulk of your security operations team, or those that are not likely to be used regularly, such as forensic security analysts.[1] Outsourcing offers access to a broader talent pool and advanced technologies that are likely too expensive or complex to maintain in-house unless you have a large, well-funded security function.

When hiring staff, a key criterion I look for is a focus on continual self-development. Information security is such a broad and ever-changing field that your skills will be rapidly outdated unless you have the interest and initiative to keep yourself up to date. You can gauge this in your potential hires by looking into their certifications, self-directed training via online resources, their home lab if applicable to the role, their memberships to information security groups, both local and national, etc. These are the people you want. What you do not want is someone who just completed a 6-week boot camp in security and now thinks that they are qualified to be a CISO because they know how to read a PCAP file dump.

You want to ensure that you have a job description describing the key responsibilities of each role you fill. Nothing is more useless than a generic information security job description that covers all the myriad aspects of security and ends with the statement, "Other duties as assigned." As the saying goes, you will end up with the staff who are "Jacks of all trades and masters of none."

These generic job descriptions typically arise when Human Resources is responsible for the hiring process, as they do not understand the diversity and complexity of the security functions. You need to fight back against these and get functionally specific job descriptions to hire specialists in your needed roles.

Once you start building your team, in addition to implementing and managing the controls that make up your program, enlist them in developing the standards and procedures for their areas of responsibility. The more they own their areas of responsibility, the better their performance will be.

In addition, make cross-training a part of your security culture. Everyone should not only know their primary job but also be capable of acting as a backup for others should the need arise (and it will). Turnover, sickness, etc., are all part of daily life, and if you operate with just one person performing each critical function in your security organization, your department is only one absence away from disaster.

8.2 Processes

When considering all the documentation required to implement and manage a security organization, you must understand the hierarchy of policies, standards, and procedures. The hierarchy can be visualized as a pyramid, with policies at the top, followed by standards, and then procedures at the base. Each level translates the abstract concepts of the higher levels into more concrete, actionable items. This structure ensures that all security measures are traceable back to the organization's strategic objectives, as outlined in the policies, and that they are implemented in a standardized and repeatable manner through standards and procedures.

Policies

Cybersecurity policies are generally approved by the highest levels of an organization's governance structure. This often includes senior executives, such as the Chief Information Officer (CIO), Chief Information Security Officer (CISO), or the Board of Directors. The approval of policies signifies an organization-wide commitment to cybersecurity and ensures that these policies have the authority to influence behavior and decisions across the entire organization. More importantly, the policies are used by regulators and auditors to confirm this commitment to security and to assess the information security organization's operations.

As I discussed in the chapter "Present Your Plan", shorter is always better when it comes to policy creation — the policies you develop should be relatively static over time. If you need to change them regularly, they are probably too specific.

Standards

Subject-matter experts, such as IT security managers or dedicated security/technology teams, usually develop standards. They are based on the strategic direction set by the policies and detail the "how" of meeting the policy objectives.

Approval for standards might come from mid- to senior-level management, such as the CIO/CTO, the CISO, or leaders underneath them, who have the technical expertise to evaluate the standards' applicability and alignment with the organization's policies. In some cases, standards may also require legal review to ensure compliance with relevant regulations.

Standards will typically be reviewed at least annually by all involved stakeholders to ensure that they still reflect what is being done to configure and manage technologies within the organization, as well as changes and updates in these technologies. Changes in technology are a constant in all organizations. Unless your standards keep up with these changes, they will soon be both useless and the cause of a pending audit finding when an auditor points out that you are not following your own standards.

Procedures

Procedures are typically created by individuals or teams with specialized knowledge of the specific tasks involved, such as system administrators, security analysts, or operations personnel. Operational managers or department supervisors, who are responsible for the areas where the procedures will be implemented, usually approve them. These approvers ensure that the procedures are practical, clear, and can be carried out effectively by their teams.

Procedures for operating all security technologies and creating all security reports and outputs are vital to the program's long-term success. Turnover in security organizations is typically a given due to job demands and competition from other organizations. Without procedures, every new employee is not only reinventing the wheel in learning what needs to be done but also introducing critical variability into the program, significantly increasing the risk of a security incident. Consistency in security operations is vital to their effectiveness.

Metrics

As you implement each of the controls in your organization through a mix of processes and technology, you also need to consider the key metrics to track to ensure that these controls are working effectively. Without metrics, you are flying blind regarding whether your efforts effectively reduce organizational risk.

Use your risk register to track and manage the risks raised by variances in meeting these control metrics. These risks will arise during the regular day-to-day operation of your information security plan, which must be prioritized against all of the other work you have to accomplish.

A common mistake in information security governance is trying to track, manage and report on strategic and tactical risks using these same methods. DON'T DO IT. You will be defeating the objectives of Heuristic Risk Management — which is just enough process to be helpful and not too much as to be confusing or burdensome.

8.3 Technologies

When evaluating the technologies needed to implement the controls dictated by your tactical risk analysis, don't make the mistake of discounting those that have already been purchased. It is a common situation where a new CISO joins an organization, sees the chaos of the legacy of multiple security technologies and decides to throw it all out and replace it with new whiz-bang solutions that they KNOW will work.

When they leave in 18 to 24 months, what will be left is yet another partially implemented solution to a needed control, and the organization's security landscape will be worse than when they joined.

Before you make that mistake, take a thorough inventory of all the security solutions within the organization as soon as you start. In my experience, it is not unusual to find multiple SIEM solutions, anti-malware solutions, and even GRC systems partially deployed and half-heartedly managed when I joined organizations over my career.

It is rare for a security technology to be totally useless. Before you spend a massive chunk of your security budget replacing these solutions, look to see if they are salvageable with proper management and full implementation. Hopefully, the money you save can be reallocated to areas where you lack effective controls, or your staffing plan has gaps.

Also, don't fall for the sales hype about silver bullet solutions that will solve all your security concerns with a single system. The more you entrust to a single vendor, the greater your dependence on that vendor will be.

This will negatively affect your budget, as you may have lost any negotiating power and your ability to take advantage of new and better solutions to emerging threats. I recommend looking for technology solutions that interoperate well together and are capable in their field rather than attempting to find the holy grail solution for all your needs in a single vendor or choosing a best-of-breed solution for each control that does not integrate well with others.

9

MANAGE YOUR RISKS

> You keep using that word. I don't think it means what you think it means.
>
> — INIGO MONTOYA, THE PRINCESS BRIDE

OPERATIONAL RISK MANAGEMENT is all about tracking and managing the day-to-day risks that arise in any information security organization. However, before I discuss how to find and manage these risks, I want to make clear how I view operational risk in the Heuristic Risk Management context.

In line with the intent of Heuristic Risk Management that risk calculations be quick, simple, and understandable, the definitions of each of these terms should be simple to understand and explain and, more importantly, agreed upon by the key management stakeholders. In a cybersecurity context, you need to confirm that you and your leadership team are all aligned on the definitions for the terms I will cover next.

9.1 Impact

Impact refers to the harm that would be caused if a particular threat materialized. It's a measure of the severity of the consequences. Depending on the systems and data affected, the impact of a cyberattack could range from minor inconvenience to significant financial loss, reputational damage, or even endangerment of human lives.

For example, when considering cybersecurity events:

- **Low:** A successful attack would cause minimal inconvenience or minor problems that can be easily remediated. There would be no significant downtime or data loss. For example, an attacker defaces a non-critical webpage that is quickly restored.
- **Medium:** An attack would result in moderate damage, such as a temporary loss of productivity or non-sensitive data exposure. There might be some financial loss or damage to reputation, but it would not threaten the organization's viability. An example could be a malware infection that disrupts certain business operations but is contained before impacting core systems.
- **High:** A successful attack would cause operational harm, such as an outage impacting key systems. It could involve stealing sensitive data and reporting to regulatory authorities or customers.

- **Severe:** This impact could be catastrophic, threatening the organization's survival or causing severe reputational harm. For instance, a ransomware attack that encrypts critical databases and freezes the organization's operations. Another example might be a sophisticated cyber-attack on critical infrastructure, like a power grid, leading to widespread outages and public safety hazards. For a public company, it may be an incident that meets the threshold of a materially impactful incident defined by SEC rules.[1]

Impact-level definitions should be reviewed by the governing body responsible for risk oversight, usually the board's risk committee. The definitions I provided above are just suggestions: use the definitions and examples that resonate with your management team. With their agreement and input, they will more readily support the subsequent risk calculations and actions needed to address the risks.

9.2 Likelihood

Likelihood is the probability that a given threat will exploit a vulnerability and cause harm. It is often based on factors like the threat activity level, the effectiveness of current security controls, and the target's attractiveness to potential attackers.

Here is an example when considering the likelihood of being infected with malware:

- **Low:** This level represents a low (<25%) probability of the risk event occurring. An organization with

robust anti-malware technology and process controls (in alignment with CIS Controls IG1 requirements) could consider this event low likelihood.
- **Moderate:** This level indicates a moderate (25-50%) probability of the risk event. An organization with anti-malware technology and process controls that do not fully meet CIS IG1 requirements could fall into this level.
- **High:** At this level, the risk event is more likely to occur than not (50-75%). An organization that has anti-malware technology and process controls in place but the controls are not operating effectively would be at this level.
- **Very High:** This level represents a high probability of the risk event (>75%). When considering the risk of being infected by malware, this organization lacks anti-malware technology and process controls; for example, no anti-malware solution is deployed or is largely ineffective.

Likelihood, I believe, should be determined by the security leader, as it usually involves the breakdown of a technical or process control. Just as with impact, the definitions for each likelihood level should be documented along with examples so there is limited ambiguity and it is easy to calculate risk.

9.3 Risk

Risk is the combination of impact and likelihood when considering a specific event. The following risk matrix figure represents this, where the risk of an event is the intersection of the assessed likelihood and impact.

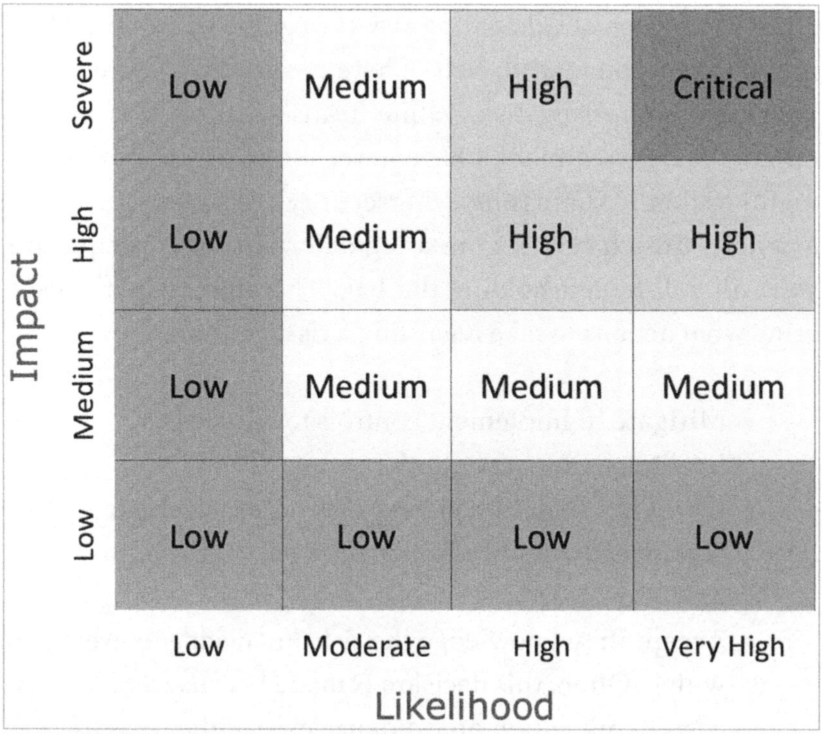

Risk Matrix

Risk, as represented in the matrix, has the following definitions:

- **Low:** The likelihood of a threat occurring is low, and/or the impact would be low.
- **Medium:** There is a moderate chance that the threat will occur and/or the impact will be medium.
- **High:** The likelihood of the threat occurring is high, and/or the impact is also high.

- **Critical:** The threat is almost certain to occur, and the impact would be severe.

The purpose of calculating risk is to guide decision-making on handling potential threats. There are generally three alternatives to addressing risk. While Transferring risk is also an option for risk treatment, I have never found this to be a meaningful option in discussing cybersecurity risks because, should a serious breach occur, from a regulatory or legal perspective, you will still be left holding the bag. Therefore, when considering what actions to take regarding a risk, you can:

- **Mitigate it:** Implement controls to reduce the likelihood, impact or both.
- **Avoid it:** Change business practices or terminate certain activities to eliminate the possibility of the risk.
- **Accept it:** Acknowledge the risk and decide to live with it. Often, this decision is made because the costs of mitigation or avoidance exceed the costs of the risk should it occur. • This is also the default for tactical risks identified as part of the risk assessment but not being actively mitigated. Since they are not being mitigated, the organization's management has de facto accepted the risks until they are addressed at a future date.

9.4 Risk Appetite

How an organization addresses risk is influenced by its risk appetite, which is the amount of risk it is willing to accept in pursuit of its objectives. Risk appetite reflects management's attitude towards risk and is shaped by the organization's culture and business priorities.

Risk and risk appetite are related in that a certain level of risk should drive decisions on handling the risk based on the organization's risk appetite. For example, a tech startup might have a high risk appetite and be willing to accept more significant risks for rapid growth and innovation. In contrast, a financial institution might have a low risk appetite due to regulatory requirements and the high impact of potential data breaches on customer trust and financial stability.

The following table presents examples of these varying risk appetites. Note that these examples are to illustrate a point—what your company adopts as its risk appetite will depend on how tolerant your leadership is of cyber risk.

Risk Level	Low Risk Appetite	Typical Risk Appetite	High Risk Appetite
Critical Risk	Address Immediately/Avoid	Address Immediately/Avoid	Address within 3 months
High Risk	Address within 3 months	Address within 6 months	Address within 1 year
Medium Risk	Address within 1 year	Add to the backlog	Add to the backlog or accept
Low Risk	Add to the backlog	Accept	Accept

Example Risk Appetite Options

Management's decision on how open they are to risk, as defined by their risk appetite, guides the development of risk management policies and procedures and, more importantly, your information security plan to address the risks uncovered. The examples I provided are just that: examples that you can use to either define the organization's risk appetite or verify it. It is less important what the actual action determinations are than it is that they are defined and everyone agrees on what they mean.

9.5 Risk Exceptions

This then brings up the issue of exceptions. No sooner are the risk policies, procedures, and definitions defined than some business executives will cry that an exception must be made for them; otherwise, the business will collapse.

A perfect example is a defined policy that no software must be released with critical flaws (as determined by the methodology used by the application security team). Of course, as soon as this is said, the sales manager will object that they must release the next version of the product by a fixed date; otherwise, a critical customer deal will be lost, and the revenue associated with the deal can make or break the company.

What to do? As part of your risk procedures, you must specify who can make an exception to the defined risk appetite requirements. A typical example would be that the board (or some person or committee of the board) approves any Critical risk exceptions. The CEO may have approval authority for High risk exceptions, and a combination of the CIO/CISO and impacted business executive may approve acceptance of Medium risks.

These exceptions must be documented in the risk register (discussed next), along with any mitigating actions to be taken and a deadline by which the risk will be expected to be mitigated.

It is important to note the difference between a risk exception and a risk acceptance. In both instances, the risk is not being mitigated. However, in the case of an exception, there is an acknowledgment that it should be mitigated according to the company's risk appetite and policy, but for whatever business reason, it has been decided that it will not be addressed at this time. This is why getting management sing-offs on these variances is so critical.

A risk acceptance is just that, an acknowledgment of the risk and the decision that nothing will be done about it. Exceptions are de facto acceptances and should be noted as such on the risk register, which I will discuss next.

9.6 Risk Register

Finally, we move on to the risk register, the primary tracking tool you will use and update to track and manage your tactical and operational risks. It functions as a comprehensive record that documents the organization's risk exposure and facilitates informed decision-making, prioritization of security tasks, and communication of risk across the organization.

The primary purpose of a risk register is to ensure that cybersecurity risks are identified and documented in a central location. Once these risks are recognized, the register is used to analyze each risk's potential impact and likelihood and prioritize them based on severity. This ranking, in combination with the risk appetite statement, determines which risks need immediate attention and resources and which can be addressed later. The risk register also outlines mitigation strategies for each risk and tracks the progress of these efforts over time.

The table below outlines the typical fields contained in a risk register. As you will note, a spreadsheet will work fine for tracking this information, and most organizations, even very large ones, typically use spreadsheets for this function. A GRC is nice to have but is not required for an effective risk register and management function. More crucial are the discipline and documentation to enforce a defined process.

Risk Register Field	Field Definition
Risk ID	A unique identifier for each risk for tracking purposes
Risk Description	A detailed explanation of the risk, including how it could impact the organization
Inherent Likelihood	How likely the risk event is to occur based on current controls
Inherent Impact	The impact of the risk event should it occur based on current controls
Inherent Risk	The calculated risk as it currently exists
Risk Decision	Mitigate, Accept or Avoid
Risk Owner	The person responsible for the Risk Decision
Risk Exception	Flag for risk exceptions
Action Plan	The planned actions to mitigate the risk
Target Date	The planned date by which the risk will be mitigated
Residual Likelihood	How likely the risk event is to occur after the mitigations are made
Residual Impact	The impact of the risk event should it occur after mitigations
Residual Risk	The calculated risk after mitigations
Status	The current status of risk mitigation, used in risk reviews
Risk Status	Pending Mitigation, Mitigation Underway, Closed (date)

Sample Risk Register Fields

When it comes to the initial risk input, especially for those who require approval or have requested an exception, a form is typically created with a subset of this information that can then be routed for review and signature (either electronically or physically). This form is then assigned the same risk ID as the one entered into the register, and it is stored for record-keeping when finalized. From that point forward, the risk is tracked in the register.

By evaluating the specific levels of impact and likelihood for each identified threat, an organization can prioritize its risks and focus its cybersecurity efforts where they are most needed. This process is crucial to risk management and helps ensure that resources are allocated effectively to protect the organization's assets.

10

FIND YOUR GAPS

> Victorious warriors win first and then go to war, while defeated warriors go to war first and then seek to win.
>
> — SUN TZU, THE ART OF WAR

OPERATIONAL RISK MANAGEMENT is all about finding the areas of your information security program where there are significant gaps in missing controls or where your existing controls are not working as intended. These issues will either highlight the need for an additional initiative, which then goes into your next revision of your information security plan (or the current plan if the gap is of such risk that it displaces initiatives you currently have underway), or they will be items that require less than a formal budgeted project but still need an action plan to address them.

Just as all ships of significant size have bilge pumps to drain the water and oil constantly accumulating at the bottom of the hull, [1] so does the operational risk register, which collects and disposes of the risk issues uncovered continuously in security programs. The absence of a risk register is akin to boating without a functioning bilge pump. Initially, everything may seem fine, but one day, you find that your ship is riding low in the water and in danger of sinking due to the accumulated water in the bilge. Similarly, unaddressed security issues can pile up until your company is breached by a collection of easily correctable issues if only you had taken the time to address them.

There are three primary ways to collect the issues that go into your risk register. These are:

- **By observation**, when the performance metrics you have established for existing controls are not met
- **By detection**, when you or an external party conducts reviews or tests of your environment or program
- **By incident**, when, as part of your incident response process, you uncover gaps in controls or failures of existing controls

10.1 By Observation

All key controls in your information security program should also have associated performance metrics that you track to ensure that they are operating as intended. These controls should align with the CIS controls you used to assess your organization.

A perfect example is the CIS Control 7 for Continuous Vulnerability Management.[2] Assuming that you have implemented an automated vulnerability scanning tool that classifies vulnerabilities by The Common Vulnerability Scoring System (CVSS) score,[3] a simple system to drive the remediation of these issues is to use the severity levels established by the PCI Security Standards Council to drive your patching efforts.[4]

The PCI severity levels are High (vulnerabilities with CVSS Scores of 7.0 to 10.0), Medium (vulnerabilities with CVSS scores of 4.0 to 6.9), and Low (vulnerabilities with CVSS scores of 0 to 3.9). If your patching program dictates that all High vulnerabilities must be patched within 30 days and all Medium vulnerabilities within 60 days, then when your regular vulnerability scanning efforts show that patching for High Vulnerabilities exceeds 30 days, you should add this to the risk register.

Why bother, you may ask, as this is not going to make the patching occur if it was not being done already? The answer is visibility. By adding it to the register, assigning it an owner, and asking the owner to set a date by which it will be done, you are making the problem visible, both to senior management and, in particular, the board.

There may be perfectly valid reasons why the patching is not occurring according to the standard. The IT organization (who should be responsible for patching) is understaffed or has competing business priorities, there is insufficient time to adequately test all the systems that the patch needs to be applied to, or there are inadequate testing facilities, the patch may break a critical application and therefore cannot be applied, etc. These are legitimate excuses for why the standard is not being met. What is crucial to understand is that without visibility in the risk register, these issues will never be addressed.

I have lost count of the environments I have walked into where systems have had unpatched high-risk vulnerabilities that stretch back years if not decades. As the saying goes, "Out of sight, out of mind." Without a defined performance metric and a process for raising the visibility of failures to meet the metric, the problems will simply not be addressed.

Raising the visibility of the problem is not a magic wand for fixing it. However, when the issue is on the register, it will hopefully drive the discussions about what actions must be taken to fix it. Does IT need additional staff to maintain the patching cadence? Do they need to adjust their priorities so that more staff is available to work on this effort? Does the application that will not allow the patch need to be upgraded or replaced? All of these are potential solutions to addressing the risk on the register.

Most importantly, there will now be a discussion of these options, and if the problem cannot be resolved, force senior management or the board to accept and create an exception for the risk. This is crucial because when the auditor or regulator examines your information security program, especially in response to a breach, you must be able to demonstrate that the risk was identified and raised and that MANAGEMENT made the business decision to either not address or postpone addressing the problem. If there is no record of this risk being raised, then the assumption will be made that you, as the CISO, are incompetent and/or your information security program is flawed in its design. Not what you want to happen!

10.2 By Detection

In addition to your control management efforts, you also need to ensure that you have regular reviews of your program and processes. Hopefully, this will be a combination of reviews by both internal and external resources.

Internal resources can be as simple as a member of the information security team tasked with the internal audit function of continuously reviewing the information security policy, standards, and procedures, ensuring that they are in place and working as intended. Even better is working with the organization's internal audit department to structure these same activities, as I will discuss in the next chapter, "Test Your Defenses".

External penetration tests (CIS Control 18 Penetration Testing) are another way to detect flaws or gaps in your security controls.[5] The objective of penetration testing is to identify weaknesses in both technical systems and human factors.

Penetration testing goes beyond just technical testing of controls and also tests the human aspect of security. Social penetration testing, for instance, assesses employees' susceptibility to social engineering tactics like phishing, pretexting, and baiting.[6] This form of testing evaluates how well individuals adhere to security protocols and whether they can be influenced to compromise security.

Physical penetration testing is another crucial aspect, where testers may employ tactics like impersonation or dumpster diving.[7] Impersonation involves testers posing as authorized personnel to gain physical or virtual access to sensitive areas, while dumpster diving involves searching through trash to find sensitive information that has been improperly disposed of. These tests are instrumental in assessing the effectiveness of physical security controls and employees' adherence to access control and information disposal policies.

Red, blue, and purple team assessments are also integral to penetration testing.[8] Red team assessments simulate real-world attacks to test how effectively an organization can defend against an aggressive breach. In contrast, blue team assessments focus on strengthening the organization's defenses and response strategies. The collaboration of these teams in purple team exercises fosters a more dynamic and responsive security posture, combining offensive insights with defensive strategies for enhanced security.

The results from all these various forms of penetration testing are invaluable for identifying weaknesses in technical controls and organizational policies and procedures. All the findings from these reports should be reviewed and risk-rated by your team and entered into the risk register with appropriate assignments and recommendations for action. Penetration testing companies, in my experience, tend to inflate the ratings of the findings that they uncover. Your job, as CISO, is to filter these findings based on your qualitative risk analysis of likelihood and impact. You do not need to accept theirs verbatim.

10.3 By Response

The last means by which risks are identified for inclusion in the risk register is through the incident response processes (CIS Control 17 Incident Response Management).[9] This can occur either pre-incident, through findings that arise through incident preparation activities such as tabletop exercises,[10] or when actual incidents occur as part of the incident investigation and lessons learned.

A structured incident response (IR) process is critical to your organization's risk management strategy.[11] Before an incident occurs, this process enables you to detect and evaluate risks for inclusion in your risk register, ensuring you're prepared for potential security events.

One of the cornerstones of an effective IR process is the incorporation of threat intelligence.[12] This involves gathering, analyzing, and applying information about current threats in your industry and business environment to anticipate and prevent potential incidents. By understanding the threat landscape using information gained from the various sources and vendors of this information, you can identify new risks or reassess existing ones in your environment. Using this information, you can identify areas where your existing controls may be deficient and thus ensure you stay ahead of your attackers.

In addition to threat intelligence, IR testing activities such as tabletop exercises are invaluable for detecting risks for inclusion in the risk register.[13] These can include everything from deficiencies in existing standards or procedures to gaps in technical controls that either do not cover all needed areas or are not working as intended. All of these findings should be recorded in your risk register, as they represent potential risks that need to be addressed through additional controls or by refining existing ones.

After an incident, post-incident analysis and lessons-learned sessions are integral to understanding what happened, why it happened, and how effectively your team responded.[14] This analysis often reveals controls that were missing, inadequate or did not function as intended during the incident. All of these are additional findings for your risk register so that they can be corrected.

~

For those familiar with more traditional risk processes, typically employed in larger organizations, you will note that one method I have not addressed for uncovering risks is the brainstorming and survey methods typically used in these organizations. The concept of group intelligence is that if you ask enough people a question regarding a matter they have little to no knowledge of and average the answers, you will somehow gain valuable insights you would not otherwise obtain, for example, by asking people who actually know the subject area.

I am sure that blind squirrels occasionally dig up random acorns; likewise, broken clocks are accurate twice daily. However, neither of these answers is sufficient justification for not focusing on the practical measures I have outlined above.

11

TEST YOUR DEFENSES

The enemy of a good plan is the dream of a perfect plan.

— PRUSSIAN GENERAL KARL VON CLAUSEWITZ

HOW WILL you know if your security program is working? Certainly, the lack of significant security incidents is one (bad) method. A better way is to regularly engage various third parties, both external and internal, in reviewing your program to provide a fresh set of eyes on what you are doing or plan to do. This can be done through external attestations, and external and internal audits.

11.1 External Attestations

In cybersecurity, an external attestation refers to an independent third-party organization's formal evaluation or certification verifying that a company's cybersecurity policies, procedures, and controls meet specific predefined standards or regulatory requirements. While this attestation may be required based on your industry, this does not mean you cannot use these attestations if you are not in that industry. Also, it does not mean that you have to wait to be completely ready, even if you are required to meet an attestation requirement such as PCI.

Most firms that perform these assessments also provide pre-assessment services where they will review your program and identify gaps to be closed before a formal assessment. This can be used to identify additional controls you can add to strengthen your cybersecurity program once you have completed your foundational CIS implementation. The most commonly used external cybersecurity attestations are:

- **ISO/IEC 27001:** A globally recognized standard for managing information security, focusing on the establishment and maintenance of an information security management system (ISMS).
- **SOC 2 (Service Organization Control 2):** Aimed primarily at technology and cloud computing organizations, this attestation ensures that service providers securely manage data to protect their clients' interests and privacy.[1]
- **PCI DSS (Payment Card Industry Data Security Standard):** This standard is essential for any

organization handling credit card transactions. It is designed to protect credit and debit card transactions from data theft and fraud.

- **CMMC (Cybersecurity Maturity Model Certification):** Specifically designed for defense contractors in the U.S., this certification is focused on protecting sensitive defense data.
- **FedRamp (Federal Risk and Authorization Management Program):** A mandatory third-party assessment of the security of cloud services used by federal agencies to confirm the effectiveness of the controls for assessing, authorizing, and continuously monitoring cloud services and products.

11.2 External Audits

In addition to engaging third-party audit and assessment firms to attest to the standards mentioned in the prior section, these firms also play a vital role in helping organizations assess their security program's maturity against any cybersecurity controls framework, not just a particular attestation. These services can provide a roadmap to achieving the eventual certification for a specific attestation, such as PCI DSS, or can be used to assess and improve the organization's overall cybersecurity posture when using a standard such as the NIST Cyber Security Framework (CSF).

Using external auditors rather than doing this in-house has a significant advantage in terms of specialized knowledge and experience. They often possess deep insights into best practices and emerging threats within specific industries. This perspective is crucial in identifying potential blind spots and vulnerabilities, enabling proactive risk identification and remediation.

Furthermore, the credibility conferred by external auditors is substantial, enhancing trust among customers, partners, regulators, and, most importantly, the board and senior management. The saying goes, "You often pay external auditors to read your watch and tell management what time it is." Frustratingly, when they do the reading, as opposed to the security leader, their statements are taken at face value, while the leader's statements are viewed skeptically. The wise security leader understands this dynamic, uses it to promote and advance their security program, and welcomes their input and reports.

11.3 Internal Audits

Internal audit functions within an organization play a pivotal role in providing independent assurance to the board regarding the organization's inherent risks, including those related to cybersecurity. This independent assurance is crucial for the board's clear and unbiased understanding of the risk landscape, enabling informed decision-making at the highest level.

The internal audit's primary function is to objectively assess and report on the effectiveness and adequacy of risk management, control, and governance processes. By evaluating the organization's cybersecurity program, internal auditors identify potential vulnerabilities, assess the likelihood and potential impact of various cybersecurity risks, and determine whether the organization's cybersecurity controls are effective and aligned with its risk appetite.

In providing this independent assurance, internal auditors support the security leader's objectives by ensuring that the board is fully aware of the organization's cybersecurity risks. This awareness is critical for garnering support and resources from top management for cybersecurity initiatives. The security leader benefits from this arrangement, as the board's understanding and backing are essential for implementing effective cybersecurity strategies.

Moreover, the internal audit's independent reports to the board can validate and reinforce the security leader's assessments and recommendations. By presenting a third-party view of cybersecurity risks and the effectiveness of current controls, internal audits can help underscore the necessity of proposed security measures or investments.

Internal auditors also play a vital role in bridging the communication gap between the security team and the board. They translate complex cybersecurity issues into business terms that the board can understand and relate to the organization's overall risk framework. This translation is critical to ensuring the board comprehensively understands the cybersecurity challenges and their potential business impacts.

When an internal audit identifies gaps, you must rate them yourself and then add them to the risk register based on your rating. This is especially important as there will often be items raised that are essentially "nits" that are low risk. However, it is a requirement of internal audit that they get a response to EVERY finding that they generate.

Put it on the risk register, and if appropriate, rate it low and accept the risk (if your risk policy and risk appetite statement allow). In this way, all items are tracked and addressed, even if you end up in lengthy discussions with the internal auditor about why you are not actioning every finding based on your risk assessment. Point them back to the risk appetite and policy, stick to your guns, and you should win the argument.

The internal audit function can be a valuable ally to the security leader if managed properly. By providing independent assurance to the board about cybersecurity risks and the effectiveness of the organization's cybersecurity program, internal auditors facilitate informed decision-making, support the security leader's objectives, and strengthen the organization's cybersecurity posture.

12

REPORT YOUR PROGRESS

> There are known knowns — there are things we know we know. We also know there are known unknowns — that is to say, we know there are some things we do not know. But there are also unknown unknowns, the ones we don't know we don't know.
>
> — U.S. SECRETARY OF DEFENSE DONALD RUMSFELD

IMPLEMENTING an information security plan and program involves more than just doing it; you must also report your progress to the various stakeholders you have. For information security, in particular, you are likely to have multiple stakeholders that you need to keep apprised of your plan and progress. Here are my recommendations for the minimum reporting meetings you should consider putting in place.

12.1 Steering Committee

First, an information security steering committee (ISSC) should be established if one is not already in place. This group should comprise the critical stakeholders for the information security function. You will want the senior leaders for these functions to attend, as the security program directly or indirectly impacts them.

Depending on your organization's size and structure, consider the following as candidates for the ISSC. However, I would caution you against inviting executives just to increase the group size. The bigger the group, the harder it will be to manage and maintain a focus on the security program and plan.

The following are who I consider the critical members of an Information Security Steering Committee:

- **Chief Information Security Officer (CISO):** The CISO is usually the committee chair and is responsible for the overall direction and management of information security within the organization.
- **Chief Information Officer (CIO):** The CIO (or Chief Technology Officer — CTO) oversees the IT department and ensures that the information security strategy is aligned with the organization's technology initiatives.
- **Chief Compliance Officer (CCO) or Legal Counsel:** This member addresses legal and regulatory compliance issues related to information security and privacy.

- **Chief Privacy Officer (CPO):** If the organization has a CPO, their involvement in the ISSC will help ensure alignment between security and privacy interests in protecting personal data and that security measures align with privacy laws and regulations.
- **Key Business Unit Leaders:** Representatives from various business units directly involved with or impacted by the security program and strategy.

The following executives are typically not involved in the ISSC: the Chief Executive Officer (CEO), Chief Operating Officer (COO), Chief Financial Officer (CFO), Chief Revenue Officer (CRO), and Chief Human Resources Officer (CHRO). While security may certainly impact all of their functions, it is unrealistic to expect them to have sufficient interest in information security to attend on a regular basis. Other executives, such as non-key business unit leaders, may attend when security concerns specific to their areas need to be discussed and addressed.

12.2 Monthly Reporting

The monthly Information Security Steering Committee meeting is a structured forum where the ISSC members come together to discuss, review, and guide the organization's information security program and plan. The agenda and purpose of such a meeting typically include the following components — tailor these as necessary to your organization:

- **Information Security Status Update:** Review the current threat landscape, highlighting any

significant changes that could impact the organization. Summarize recent security incidents (if any) and their status.
- **Incident Analysis and Lessons Learned:** Discuss significant security incidents since the last meeting, including root cause analysis.
- **Policy and Compliance Updates:** Discuss upcoming regulatory changes, if any, and their potential impact.
- **Risk Management Updates:** Present recent risk assessments, audits or penetration tests, and findings. Review the risk register, review new items, and assess progress against existing risks.
- **Project and Initiative Updates:** Status updates on ongoing security projects and initiatives.
- **Training and Awareness Programs:** Review of the status and effectiveness of security training and awareness programs. Review of ongoing security testing for staff, such as monthly phishing testing.
- **Metrics and Reporting:** Review security metrics and key performance indicators (KPIs) for key security controls, including discussion of actions to be taken on any controls not operating within defined parameters.

Because you are presenting to other executives, keep the slides and content concise and focused. I recommend that all metrics define the acceptable ranges for each and use a Red/Yellow/Green stoplight to highlight problem areas. Don't spend much time discussing items that are operating within defined parameters; focus on the problem areas and the actions you plan to take to address the variances (or that other executives, such as the CIO, will take if they are responsible for a control that is operating outside of defined parameters, such as patching).

Likewise, remember that the goal of the meeting is not only to keep key stakeholders aware of plans and progress but also to gain their support for overcoming challenges you may be facing that are slowing progress or adoption of the information security plan. There will always be something.

I also suggest tracking and presenting control metrics over time where applicable, such as phishing testing results, so that within a year, you will present a year's worth of data. When presenting this information, show it as a line chart and track and present a trend line along with the actual data so that you can monitor and track the trend, which is what really matters.

Likewise, when establishing the parameters of what constitutes Red, Yellow or Green when presenting each area, consider using Yellow to indicate a one-month change from Green (within defined parameters) and don't immediately change it to Red (unless there is a reason to highlight an urgent change, such as a significant rise in the threat level due to targeted attacks against your organization). Only change the indicator to Red in the 2nd month to indicate a continuing trend. This is because numerical metrics tend to bounce around, and you don't want to spend much time discussing an issue that could be nothing more than statistical noise.

12.3 Quarterly Reporting

A quarterly risk review between the Chief Information Security Officer (CISO) and the board's risk committee (or the CEO if the board is not involved in cybersecurity) is an essential meeting that focuses on evaluating and managing cybersecurity risks within an organization. The agenda and purpose of such a meeting typically include the following key components — tailor these as necessary to your organization:

- **Review of the Current Risk Landscape:** Present the organization's current cybersecurity threats and vulnerabilities. Discuss any new types of cyberattacks or threat actors that have emerged since the last meeting. Discuss the impact of external factors such as regulatory changes, technological advancements, or global cybersecurity trends.
- **Performance Metrics and Risk Assessment Updates:** Review key performance indicators (KPIs) and metrics that track the effectiveness of the cybersecurity program. Provide an update on the status of risk assessments, including any changes in the organization's risk profile. Assess the effectiveness of current risk mitigation strategies and controls.
- **Incident Report Analysis:** Review security incidents that occurred during the quarter, including breaches, data leaks, and attempted attacks. Discuss the incident response process and any lessons learned. Discuss any necessary changes to incident response plans or procedures.
- **Compliance and Regulatory Updates:** Provide an overview of compliance obligations related to cybersecurity and any upcoming regulatory changes. Provide a status update on compliance efforts, any identified gaps or issues, and plans for addressing compliance requirements while maintaining adherence to relevant laws and standards.

- **Budget and Resource Allocation:** Review the cybersecurity budget, including current expenditures and forecasted needs.
- **Strategic Initiatives and Roadmap:** Review strategic cybersecurity initiatives and their progress toward long-term goals. Discuss the cybersecurity roadmap and any adjustments needed based on the evolving risk landscape.
- **Stakeholder Communication and Education:** Discuss strategies for communicating risk-related information to stakeholders, including employees, customers, and partners. Discuss plans for ongoing cybersecurity awareness and training programs within the organization.
- **Executive and Board Engagement:** Discuss maintaining active engagement with executive leadership and the board regarding cybersecurity matters. Review any feedback or concerns expressed by board members or executives.
- **Action Items and Follow-Up:** Identify action items and assign responsibilities to address any issues raised during the meeting. Determine timelines and deadlines for completing action items. Plan for the follow-up on outstanding issues and report on progress at the next meeting.

The purpose of these meetings is to ensure that the board's risk committee is fully informed about the current state of cybersecurity within the organization, that strategic decisions are made based on the latest risk intelligence, and that there is alignment between the CISO's activities and the board's risk appetite and business objectives. It also demonstrates accountability and due diligence in managing cybersecurity risks, which is critical for the organization's reputation and stakeholder trust, particularly to regulators who will want to see the minutes of these meetings to ensure the board's active engagement in monitoring risk.

12.4 Annual Reporting

As discussed in the chapter "Present Your Plan", you presented an abbreviated version of the materials usually covered in the annual information security report to the board in your initial funding request for your cybersecurity program. When it comes time for your first formal yearly update to the board, you will want to be prepared to cover all the items specified under the Final Rule, as defined in 6 CFR 314.4(i). Using this as a guide, consider the following agenda for your annual report to the board, tailored as they see fit or are used to seeing:

- **Overall Status of the Information Security Program:** Provide an executive summary of the program's current state, highlighting any significant changes or developments since the last report. Outline the program's alignment with the institution's objectives and regulatory requirements.
- **Compliance with the GLBA Safeguards Rule:** Detail the institution's efforts to comply with the GLBA Safeguards Rule (or any other regulatory requirement for information security applicable to the organization), including any compliance assessments or audits conducted. Discuss any areas of non-compliance, plans for remediation, and progress on previously identified issues.

- **Risk Assessment Results:** Present the most recent tactical risk assessment results, including areas of weakness identified. Review the methodology used for the risk assessment and any changes from previous assessments.
- **Risk Management and Control Decisions:** Summarize the risk management strategies and control decisions made in response to the risk assessment findings. Discuss any changes to existing controls or the implementation of new controls.
- **Service Provider Arrangements:** Review the oversight and management of third-party service providers, including how their compliance with information security requirements is assessed and monitored. Highlight any changes to service provider arrangements that may impact the information security program.
- **Results of Testing:** Report on any testing of the information security program, such as penetration testing, vulnerability assessments, or audits. Discuss any identified weaknesses and the steps taken to address them.
- **Security Events or Violations and Management's Responses:** Detail any security events or policy violations that have occurred, their impact, and management's response to them. Discuss any lessons learned, as well as how they have been or will be integrated into the information security program.
- **Recommendations for Changes in the Information Security Program:** Based on the outcomes of risk assessments, testing, and

monitoring activities, provide recommendations for improving the information security program. Outline any proposed changes to policies or controls, including the rationale behind these recommendations, and seek their approval.

- **Associated Budget for the Information Security Department:** Present the current budget and financial resources allocated to the program. Propose any budgetary adjustments or additional funding required to implement recommended changes and address identified risks.
- **Action Plan:** Include an action plan that outlines the steps and timelines for implementing the recommendations and addressing any identified issues. Assign responsibilities for each action item to ensure accountability.

The board should review and discuss these items with the CISO to ensure they have a comprehensive understanding of the information security program's status and the actions required to maintain compliance with the GLBA Safeguards Rule (or whatever other regulation applies to their information security oversight responsibilities). The board's approval of the report and any proposed changes to the information security program is critical to demonstrating oversight and their commitment to safeguarding customer information. The meeting should be well-documented, and follow-up actions should be tracked to ensure the recommendations are effectively implemented. This is also another critical item that regulators will be looking to review.

12.5 The Forever War

So, to recap all we have covered, Heuristic Risk Management is an approach to developing and managing an information security program that uses risk to prioritize action.

At the strategic level, the risks are the threats posed by your enemies. At the tactical level, the risks are the gaps in your ability to defend against your enemy's attacks. At the operational level, the risks are related to your ability to implement and operate these defensive measures effectively. Understanding these risks, as well as the processes for determining and managing them, is what separates effective information security programs from pending disasters.

It is also important to stress that the information security battle is never-ending. This is truly a "forever war", as there is no truce, armistice or victory at the end. There is only the possibility of defeat if you are unlucky or have not built effective defenses.

As you determine and implement your defenses, your enemies are evolving their attacks to bypass them and find new ways to achieve their aims. Just like the castles of yesteryear, the information security program that is effective today will be ineffective in the future if its success is defined by its static defenses.

It is the process of continually building and evolving your defenses based on the risks you face (which I hope you have learned from this book) that is your only chance for survival in today's cyber battlefield. My best wishes for your success...

— Michael Lines

Be Aware, Get Prepared, Defend Yourself

AFTERWORD

> It must be remembered that there is nothing more difficult to plan, more doubtful of success, nor more dangerous to manage than a new system. For the initiator has the enmity of all who would profit by the preservation of the old institution and merely lukewarm defenders in those who gain by the new ones.
>
> — NICCOLÒ MACHIAVELLI

There is an elephant in the room whenever a new CISO joins an organization — one that most do not see until it is too late and sitting on their chest. That elephant is the CISO's misperception that the people they work with and for have the same desires, knowledge, and incentives as they do and that they all want to do the "right" thing. All they need is guidance from the CISO on what those "right" things are. Right? Wrong!

This is an easy mistake to make, especially when the company's commitment to maintaining the security and privacy of its customers' information is touted on its website, public proclamations, and marketing materials. However, businesses are no different than people (because, of course, people make them work) in that what people say and actually believe and do are often highly divergent.

Unless and until a CISO can gain at least the top-level support (and it must start with the Board and/or the CEO) for the security program and objectives, they will likely be doing nothing more than treading water when it comes to addressing the organization's security needs.[1] That top-level support, publicly stated and demonstrated, is what is required to support the CISO in overcoming the biases that exist at all levels of the organization. These biases exist for:

- **The Board**: Board directors are often as concerned with maintaining their positions as they are with the welfare of the organization. They are generally not technically proficient or actively involved in security plans or risks. The lack of direct, honest, and ongoing discussions with security leaders compounds their lack of knowledge. What information they receive regarding the organization's security posture is often censored or filtered by the management chain through which security reports. This usually happens to ensure that the reports from the security leader do not contradict prior assertions that they may have already received from management about the environment's health.
- **CEO**: The CEO is usually focused on driving the organization to meet its business objectives. The

security function can be seen as a risk and a potential blocker to these plans. Normalcy Bias plays into this as well.[2] No serious security incident has happened in the past, and therefore, we must have good security, and nothing bad will happen to us in the future.

- **Information Technology Leaders:** The CIO/CTO often have different priorities and incentives than security leaders — uptime, for example — and are resistant to security recommendations, such as patching, that they see as putting their objectives at risk. Also playing into this is the difficulty new security leaders have after joining an organization, where they are likely to point out security flaws that IT has either failed to address or said were fixed when they were not. Normalcy Bias can play a significant role with these leaders as well.
- **Your Business Peers:** For those leaders responsible for production and revenues, their incentives are aligned around meeting specific goals in these areas. Security can be seen as a blocker to these goals by imposing requirements such as contract reviews, vendor risk analysis, and design reviews, all of which lead to delays and costs in these leaders' plans.
- **Your Organization's Employees:** Finally, everyone else. Employees are generally interested in doing their jobs with the least friction and effort. As a result, they resist controls that require them to jump through hoops or impose controls they do not have in their personal lives, like site restrictions, device restrictions, and timeouts. Since they don't understand why these controls exist, they will often

> bypass them if they impede their productivity (such as sending email/files to home addresses, sharing files on unencrypted devices), make stupid mistakes (leaving laptops exposed, clicking on any interesting email or attachments), and finally attempting to cover up or not report their mistakes until it is too late.

This all boils down to a lack of awareness within the organization of the real risks posed by cybersecurity threats and an unwillingness to compromise to mitigate them. Therefore, the best time for a CISO to join any company is after they have had a major public breach with significant regulatory and legal sanctions. There is nothing like being sued and putting your company's life at risk to get everyone's attention that this security stuff might really matter. Conversely, the worst time to join (or stay) is when the company is in denial of its risks and is unwilling to make the needed changes and investments. In that case, the CISO is just the patsy waiting to be thrown under the bus when the inevitable happens.

What the security leader must do, especially with the board and senior management, is to educate them that the worst that can happen is not the event; it is the failure to anticipate and prepare for it. The real dangers are either the cover-up of the actual situation or the active ignorance of the situation. Even if the worst happens, the incident is often survivable if you can demonstrate that you were aware, prepared, and responded to the best of your ability.

Here, then, is the real job and challenge for a CISO — you are not (just) there as a technology and process expert; you are there as a risk-aware cultural change agent. To successfully secure an organization, a CISO must overcome all of the internal biases, blinders, and incentive mismatches mentioned above.

Do you have it in you? That will be for you to decide after you have gone through the HRM process to understand the threats, assets, vulnerabilities, and risks to the organization. If you cannot get support for the plan you develop to address these issues, do you want to be the one holding the bag when the inevitable occurs? The choice is yours...

> It is difficult to get a man to understand something when his salary depends on his not understanding it.
>
> — UPTON SINCLAIR

NOTES

Introduction

1. *Weak security controls and practices routinely exploited for initial access.* (n.d.). Cybersecurity and Infrastructure Security Agency CISA. Retrieved April 23, 2024, from https://www.cisa.gov/news-events/cybersecurity-advisories/aa22-137a
2. *The importance and effectiveness of cyber risk quantification.* (n.d.). Fairinstitute.org. Retrieved April 23, 2024, from https://www.fairinstitute.org/what-is-fair
3. *Introducing OCTAVE Allegro: Improving the information security risk assessment process.* (n.d.). SEI Digital Library. Retrieved April 23, 2024, from https://insights.sei.cmu.edu/library/introducing-octave-allegro-improving-the-information-security-risk-assessment-process/
4. *About the RMF - NIST risk management framework.* (2016, November 30). CSRC | NIST. https://csrc.nist.gov/projects/risk-management/about-rmf
5. *Cyber regulation has been tried and it doesn't work.* (n.d.). Isalliance.org. Retrieved April 23, 2024, from https://isalliance.org/cyber-regulation-has-been-tried-and-it-doesnt-work/
6. Rosenbaum, E. (2021, August 10). *Main Street overconfidence: America's small businesses aren't worried about hacking.* CNBC. https://www.cnbc.com/2021/08/10/main-street-overconfidence-small-businesses-dont-worry-about-hacking.html
7. Decker, A. (2023, October 24). *What are the NIST Special Publications (SPs) 800 Series?* IPKeys. https://ipkeys.com/blog/nist-special-publications/
8. *Program basics.* (n.d.). Fedramp.gov. Retrieved April 23, 2024, from https://www.fedramp.gov/program-basics/
9. *Standards.* (2022, April 14). PCI Security Standards Council. https://www.pcisecuritystandards.org/standards/
10. *ISO/IEC 27000 family.* (2023). ISO. https://www.iso.org/standard/iso-iec-27000-family
11. Awati, R., & Cole, B. (2022, March 2). *North American electric reliability corporation critical infrastructure protection (NERC CIP).* Security; TechTarget. https://www.techtarget.com/searchsecurity/definition/North-American-Electric-Reliability-Corporation-Critical-Infrastructure-Protection-NERC-CIP
12. Goldstein, P. (2021, August 3). *What is the cybersecurity maturity model certification (CMMC) and how can it be achieved?* FedTech. https://fedtech

magazine.com/article/2021/08/what-cybersecurity-maturity-model-certification-and-how-can-it-be-achieved-perfcon

13. Hashemi-Pour, C., & Lewis, S. (2024, January 17). *OODA loop*. CIO; TechTarget. https://www.techtarget.com/searchcio/definition/OODA-loop
14. Fadilpašić, S. (2023, December 6). *This huge Russian phishing campaign is hitting targets across the world*. TechRadar Pro. https://www.techradar.com/pro/security/this-huge-russian-phishing-campaign-is-hitting-targets-across-the-world

1. Know Your Enemy

1. BBC News. (2021, May 30). NSA spying row: Denmark accused of helping US spy on European officials. *BBC*. https://www.bbc.com/news/world-europe-57302806
2. Bowen, A. S. (n.d.). *Russian Cyber Units*. Retrieved April 23, 2024, from https://crsreports.congress.gov/product/pdf/IF/IF11718
3. Laje, D. (2023, September 15). *Ukraine's fusion of cyber and kinetic warfare: Illia vitiuk's stand against Russian cyber operations*. AFCEA International. https://www.afcea.org/signal-media/test-signal-landing-page-format/ukraines-fusion-cyber-and-kinetic-warfare-illia
4. Hendery, S. (2023, November 27). *North Korean supply chain attacks 'growing in sophistication and volume.'* SC Media. https://www.scmagazine.com/news/north-korean-supply-chain-attacks-growing-in-sophistication-and-volume
5. Office of the Director of National Intelligence. (2023, February 6). *Annual Threat Assessment of the U.S. Intelligence Community*. Dni.gov. https://www.dni.gov/files/ODNI/documents/assessments/ATA-2023-Unclassified-Report.pdf
6. Tucker, P. (2021, May 28). *Russia's latest hack shows how useful 'criminal groups' are to the Kremlin*. Defense One. https://www.defenseone.com/technology/2021/05/russias-latest-hack-shows-how-useful-criminal-groups-are-kremlin/174401
7. *Nation-state cyber actors*. (n.d.). Cisa.gov. Retrieved April 23, 2024, from https://www.cisa.gov/topics/cyber-threats-and-advisories/nation-state-cyber-actors
8. *Chinese Government Poses "Broad and Unrelenting" Threat to U.S. Critical Infrastructure, FBI Director Says*. (2024, April 18). Fbi.gov. https://www.fbi.gov/news/stories/chinese-government-poses-broad-and-unrelenting-threat-to-u-s-critical-infrastructure-fbi-director-says
9. Fruhlinger, J. (2020, February 12). *The OPM hack explained: Bad security practices meet China's Captain America*. CSO Online. https://www.csoonline.com/article/566509/the-opm-hack-explained-bad-security-practices-meet-chinas-captain-america.html

10. Fruhlinger, J. (2020, February 12). *Equifax data breach FAQ: What happened, who was affected, what was the impact?* CSO Online. https://www.csoonline.com/article/567833/equifax-data-breach-faq-what-happened-who-was-affected-what-was-the-impact.html
11. Martin, A. (2023, March 31). *More evidence links 3CX supply-chain attack to North Korean hacking group*. Therecord.Media. https://therecord.media/3cx-attack-north-korea-lazarus-group
12. *Cyber Safety Review Board releases report on Microsoft online exchange incident from Summer 2023*. (2024, April 2). U.S. Department of Homeland Security. https://www.dhs.gov/news/2024/04/02/cyber-safety-review-board-releases-report-microsoft-online-exchange-incident-summer
13. Zurier, S. (2023, September 19). *Vast majority of bot attacks emanate from China and Russia*. SC Media. https://www.scmagazine.com/news/vast-majority-of-bot-attacks-emanate-from-china-and-russia
14. Porolli, M. (2019, January 31). *Cybercrime black markets: Dark web services and their prices*. Welivesecurity.com. https://www.welivesecurity.com/2019/01/31/cybercrime-black-markets-dark-web-services-and-prices
15. Kaspersky. (2018, May 25). *Malware-as-a-service (MaaS)*. Kaspersky. https://encyclopedia.kaspersky.com/glossary/malware-as-a-service-maas
16. Cimpanu, C. (2016, November 24). *You can now rent a Mirai botnet of 400,000 bots*. BleepingComputer. https://www.bleepingcomputer.com/news/security/you-can-now-rent-a-mirai-botnet-of-400-000-bots
17. *Cyber money laundering: An in-depth analysis*. (2019, July 25). Tookitaki.com. https://www.tookitaki.com/blog/cyber-laundering-cyberterrorism
18. Kassner, M. (2015, February 2). *Anatomy of the Target data breach: Missed opportunities and lessons learned*. ZDNET. https://www.zdnet.com/article/anatomy-of-the-target-data-breach-missed-opportunities-and-lessons-learned
19. *TrickBot malware*. (2021, May 20). Cybersecurity and Infrastructure Security Agency CISA. https://www.cisa.gov/news-events/cybersecurity-advisories/aa21-076a
20. Suderman, A. (2021, May 9). *Ransomware gangs get more aggressive against law enforcement*. AP News. https://apnews.com/article/ransomware-gangs-hacking-police-cybercrime-pipeline-3a38c27c4fafe0c39461fb71bf91a42a
21. *Handling threats from disgruntled employees*. (2015, July 15). SEI Blog. https://insights.sei.cmu.edu/blog/handling-threats-from-disgruntled-employees
22. Davis, J. (2022, July 27). *75% of insider cyber attacks are the work of disgruntled ex-employees: Report*. Informationweek.com; Information Week. https://www.informationweek.com/cyber-resilience/75-of-insider-cyber-attacks-are-the-work-of-disgruntled-ex-employees-report
23. Kolodny, L. (2018, June 18). *Elon Musk emails employees about "extensive and damaging sabotage" by employee*. CNBC. https://www.cnbc.com/2018/06/18/elon-musk-email-employee-conducted-extensive-and-damaging-sabotage.html

24. Brook, C. (2021, February 17). *Suit claims attorneys stole, destroyed data before joining rival firm*. Digital Guardian. https://www.digitalguardian.com/blog/suit-claims-attorneys-stole-destroyed-data-joining-rival-firm
25. Discher, E. (2017, February 16). Green: Hacks cost fired Georgia-Pacific worker years in prison, $1.1 million in restitution. *Advocate*. https://www.theadvocate.com/batonrouge/news/courts/green-hacks-cost-fired-georgia-pacific-worker-years-in-prison-1-1-million-in-restitution/article7f6ea818-f488-11e6-bada-eb1757011f89.html
26. Tucker, E. (2024, March 6). *Ex-Google engineer charged with stealing AI trade secrets while working with Chinese companies*. AP News. https://apnews.com/article/china-google-justice-department-63156ade1e564d15d92adbe f91e9c5da
27. Wikipedia contributors. (2024b, April 16). *Anonymous (hacker group)*. Wikipedia, The Free Encyclopedia. https://en.wikipedia.org/w/index.php?title=Anonymous*(hacker*group)&oldid=1219168277
28. Wikipedia contributors. (2024, February 27). *Hacktivism*. Wikipedia, The Free Encyclopedia. https://en.wikipedia.org/w/index.php?title=Hack tivism&oldid=1210589288
29. Purtill, J. (2022, February 25). Hacker collective Anonymous declares "cyber war" against Russia, disables state news website. *ABC News*. https://www.abc.net.au/news/science/2022-02-25/hacker-collective-anonymous-declares-cyber-war-against-russia/100861160
30. Molloy, D., & Tidy, J. (2020, June 1). George Floyd: Anonymous hackers re-emerge amid US unrest. *BBC*. https://www.bbc.com/news/technology-52879000
31. Krebs, B. (2022, July 26). *A retrospective on the 2015 Ashley Madison breach*. Krebsonsecurity.com. https://krebsonsecurity.com/2022/07/a-retrospective-on-the-2015-ashley-madison-breach
32. BBC News. (2011, December 26). "Anonymous" hackers hit US security firm Stratfor. *BBC*. https://www.bbc.com/news/world-us-canada-16330396
33. Protalinski, E. (2012, March 10). *Anonymous hacks abortion clinic, steals 10,000 records*. ZDNET. https://www.zdnet.com/article/anonymous-hacks-abor tion-clinic-steals-10000-records
34. Diamond, J. (2015, July 30). Planned Parenthood website hacked. *CNN*. https://www.cnn.com/2015/07/30/politics/planned-parenthood-website-hacked/index.html
35. Eliaçık, E. (2022, June 17). *What is considered cyberterrorism and why does it matter today?* Dataconomy Media. https://dataconomy.com/2022/06/17/what-is-cyberterrorism
36. *IRGC-affiliated cyber actors exploit PLCs in multiple sectors, including U.s. water and wastewater systems facilities*. (2023, December 1). Cybersecurity and Infrastructure Security Agency CISA. https://www.cisa.gov/news-events/cybersecurity-advisories/aa23-335a

37. Kovacs, E. (2023, December 8). *Cyberattack on Irish Utility Cuts Off Water Supply for Two Days*. Securityweek.com. https://www.securityweek.com/cyberattack-on-irish-utility-cuts-off-water-supply-for-two-days

2. Know Yourself

1. Travis, P. (n.d.). *How To Survive A High Intensity Firefight*. Offthegridnews.-com. Retrieved April 23, 2024, from https://www.offthegridnews.com/self-defense/how-to-survive-a-high-intensity-firefight
2. Rosencrance, L. (2023, February 15). *5 biggest risks of using third-party services providers*. CSO Online. https://www.csoonline.com/article/574543/5-major-risks-third-party-services-may-bring-along-with-them.html

3. Understand Their Threats

1. *Ransomware 101*. (n.d.). Cybersecurity and Infrastructure Security Agency CISA. Retrieved April 23, 2024, from https://www.cisa.gov/stopransomware/ransomware-101
2. *Double extortion ransomware*. (2022, October 20). Cohesity. https://www.cohesity.com/glossary/double-extortion-ransomware
3. *Understanding denial-of-service attacks*. (2021, February 1). Cybersecurity and Infrastructure Security Agency CISA. https://www.cisa.gov/news-events/news/understanding-denial-service-attacks
4. Kerner, S. M. (2023, September 13). *Triple extortion ransomware*. Security; TechTarget. https://www.techtarget.com/searchsecurity/definition/triple-extortion-ransomware
5. *2024 Data Breach Investigations Report*. (n.d.). Verizon Business. Retrieved April 23, 2024, from https://www.verizon.com/business/resources/reports/dbir
6. Mahendru, P. (2023, August 1). *The state of ransomware in state and local government 2023*. Sophos News. https://news.sophos.com/en-us/2023/08/01/the-state-of-ransomware-in-state-and-local-government-2023
7. SentinelOne. (2020, September 17). *How ransomware attacks are threatening our critical infrastructure*. SentinelOne. https://www.sentinelone.com/blog/how-ransomware-attacks-are-threatening-our-critical-infrastructure
8. Yeap, Y. P. (2021, April 30). *Why ransomware costs businesses much more than money*. Forbes. https://www.forbes.com/sites/forbestechcouncil/2021/04/30/why-ransomware-costs-businesses-much-more-than-money
9. Zacharakos, A. (2023, May 3). *Studies show ransomware has already caused patient deaths*. Security; TechTarget. http://www.techtarget.com/searchsecurity/feature/Studies-show-ransomware-has-already-caused-patient-deaths
10. Bajak, F. (2021, May 10). *Insurer AXA to stop paying for ransomware crime*

payments in France. Insurance Journal. https://www.insurancejournal.com/news/international/2021/05/09/613255.htm

11. U.S. Government Accountability Office. (2022, July 19). *Rising cyberthreats increase cyber insurance premiums while reducing availability*. Gao.gov. https://www.gao.gov/blog/rising-cyberthreats-increase-cyber-insurance-premiums-while-reducing-availability
12. *Internet Crime Complaint Center(IC3)*. (n.d.). Ic3.gov. Retrieved April 23, 2024, from https://www.ic3.gov/Home/AnnualReports
13. *Lazio Phishing Scam Nets Cybercriminals €2 million*. (2018, April 6). SpamTitan. https://www.spamtitan.com/blog/lazio-phishing-scam-nets-cybercriminals-e2-million
14. Eng, J. (2015, August 7). *Ubiquiti Networks says it was victim of $47 million cyber scam*. NBC News. https://www.nbcnews.com/tech/security/ubiquiti-networks-says-it-was-victim-47-million-cyber-scam-n406201
15. *Fraud Alert: COVID-19 Scams*. (n.d.). Hhs.gov. Retrieved April 23, 2024, from https://oig.hhs.gov/fraud/consumer-alerts/fraud-alert-covid-19-scams
16. *CEO fraud*. (n.d.). Barclayscorporate.com. Retrieved April 23, 2024, from https://www.barclayscorporate.com/insights/fraud-protection/ceo-fraud
17. *Payment Diversion Fraud: What does it look like in practice*. (2022, May 15). James Hyland & Company. https://www.jhyland.com/payment-diversion-fraud-practice
18. Rykun, E. (2022, April 8). *Payroll fraud: guide to recognition and prevention*. Spendesk. https://www.spendesk.com/en-eu/blog/payroll-fraud
19. Roberts, J. J. (2017, April 27). *Exclusive: Facebook and Google Were Victims of $100M Payment Scam*. Fortune.com. https://fortune.com/2017/04/27/facebook-google-rimasauskas
20. *Tutorial 1: The impact of cybercrime on small business*. (n.d.). Sbir.gov. Retrieved April 23, 2024, from https://www.sbir.gov/tutorials/cyber-security/tutorial-1
21. PricewaterhouseCoopers. (n.d.). *PwC's global economic crime and Fraud Survey 2022*. PwC. Retrieved April 23, 2024, from https://www.pwc.com/gx/en/services/forensics/economic-crime-survey.html
22. Vijayan, J. (2021, December 28). *4 ways employee home networks and smart devices change your threat model*. CSO Online. https://www.csoonline.com/article/569489/4-ways-employee-home-networks-and-smart-devices-change-your-threat-model.html
23. Belanger, A. (2023, March 6). *Thousands scammed by AI voices mimicking loved ones in emergencies*. Ars Technica. https://arstechnica.com/tech-policy/2023/03/rising-scams-use-ai-to-mimic-voices-of-loved-ones-in-financial-distress/
24. Edwards, B. (2024, February 5). *Deepfake scammer walks off with $25 million in first-of-its-kind AI heist*. Ars Technica. https://arstechnica.com/information-technology/2024/02/deepfake-scammer-walks-off-with-25-million-in-first-of-its-kind-ai-heist/

25. Fruhlinger, J. (2020, February 12). *Marriott data breach FAQ: How did it happen and what was the impact?* CSO Online. https://www.csoonline.com/article/567795/marriott-data-breach-faq-how-did-it-happen-and-what-was-the-impact.html
26. *Chinese company sinovel wind group convicted of theft of trade secrets.* (2018, January 24). Justice.gov. https://www.justice.gov/opa/pr/chinese-company-sinovel-wind-group-convicted-theft-trade-secrets
27. Statt, N. (2020, August 4). *Former Google exec Anthony Levandowski sentenced to 18 months for stealing self-driving car secrets.* The Verge. https://www.theverge.com/2020/8/4/21354906/anthony-levandowski-waymo-uber-lawsuit-sentence-18-months-prison-lawsuit
28. Bocetta, S. (n.d.). *How hackers use your computer to cheat at crypto mining.* Msp360.com. Retrieved April 23, 2024, from https://www.msp360.com/resources/blog/cryptojacking-how-to-detect-and-avoid
29. O'Neill, P. H. (2018, January 31). *Monero mining botnet 'Smominru' earns hackers $3.6 million.* CyberScoop. https://cyberscoop.com/monero-botnet-smominru-proofpoint
30. *'Kill Chain' Analysis of Target Data Breach is a chilling read for corporate cybersecurity and privacy professionals - Wiggin and Dana LLP — Attorneys at Law.* (2019, September 15). Wiggin and Dana LLP — Attorneys At Law; Wiggin and Dana LLP. https://www.wiggin.com/publication/kill-chain-analysis-of-target-data-breach-is-a-chilling-read-for-corporate-cybersecurity-and-privacy-professionals
31. McLean, R. (2019, July 30). A hacker gained access to 100 million Capital One credit card applications and accounts. *CNN.* https://www.cnn.com/2019/07/29/business/capital-one-data-breach/index.html
32. *How to combat mobile device perils in Healthcare.* (2017, July 8). ICTpost. http://ictpost.com/how-to-combat-mobile-device-perils-in-healthcare-2
33. *Nation-state cyber actors.* (n.d.). Cisa.gov. Retrieved April 23, 2024, from https://www.cisa.gov/topics/cyber-threats-and-advisories/advanced-persistent-threats-and-nation-state-actors
34. Muncaster, P. (2023, July 19). *NCA: Nation states using cybercrime groups as proxies.* Infosecurity Magazine. https://www.infosecurity-magazine.com/news/nca-nation-states-cybercrime
35. *North Korean regime-backed programmer charged with conspiracy to conduct multiple cyber attacks and intrusions.* (2018, September 6). Justice.gov. https://www.justice.gov/opa/pr/north-korean-regime-backed-programmer-charged-conspiracy-conduct-multiple-cyber-attacks-and
36. Pagliery, J. (2015, August 5). *The inside story of the biggest hack in history.* CNNMoney. https://money.cnn.com/2015/08/05/technology/aramco-hack/index.html
37. Sandworm disrupts power in Ukraine using a novel attack against operational technology. (2021, October 3). *Mandiant.* https://www.mandiant.com/

resources/blog/sandworm-disrupts-power-ukraine-operational-technology
38. Lyons, J. (2024, February 1). *Congress told how Chinese goons plan to incite "societal chaos" in the US*. The Register. https://www.theregister.com/2024/02/01/china*attack*warning
39. Microsoft Threat Intelligence. (2023, May 24). *Volt Typhoon targets US critical infrastructure with living-off-the-land techniques*. Microsoft Security Blog. https://www.microsoft.com/en-us/security/blog/2023/05/24/volt-typhoon-targets-us-critical-infrastructure-with-living-off-the-land-techniques
40. Goodin, D. (2024, January 31). *Chinese malware removed from SOHO routers after FBI issues covert commands*. Ars Technica. https://arstechnica.com/security/2024/01/chinese-malware-removed-from-soho-routers-after-fbi-issues-covert-commands

4. Know Their Methods

1. Dholakiya, P. (2021, March 22). *What Is the Cyber Kill Chain and How It Can Protect Against Attacks*. Computer.org. http://www.computer.org/publications/tech-news/trends/what-is-the-cyber-kill-chain-and-how-it-can-protect-against-attacks
2. *What Is Phishing?* (2021, February 18). Proofpoint. https://www.proofpoint.com/us/threat-reference/phishing
3. *What Is Vishing?* (2021, February 26). Proofpoint. https://www.proofpoint.com/us/threat-reference/vishing
4. *What Is Smishing?* (2021, February 26). Proofpoint. https://www.proofpoint.com/us/threat-reference/smishing
5. O'Donnell, L. (2020, August 27). *Malicious Attachments Remain a Cybercriminal Threat Vector Favorite*. Threatpost. https://threatpost.com/malicious-attachments-remain-a-cybercriminal-threat-vector-favorite/158631/
6. Barrus, R. (2018, October 9). *Credential Harvesting: It's More Than Just Phishing and More Common Than Ever*. Pivot Point Security. https://www.pivotpointsecurity.com/credential-harvesting-more-than-password-phishing/
7. Masas, R., Yohann, Omri, Levy, M., Sofia Naer, Johnston, D., Hasson, E., Gabi Stapel, & Cohen, O. (n.d.). *Social engineering*. Learning Center; Imperva Inc. Retrieved April 23, 2024, from https://www.imperva.com/learn/application-security/social-engineering-attack
8. *Data breaches by third parties*. (2020, September 8). Black Kite. https://blackkite.com/data-breaches-caused-by-third-parties
9. McAfee. (2022, October 20). *What are tailgating attacks and how to protect yourself from them*. McAfee Blog; McAfee. https://www.mcafee.com/blogs/internet-security/what-are-tailgating-attacks
10. McMillan, R. (2012, March 3). *The Pwn Plug is a little white box that can hack*

your network. Ars Technica. https://arstechnica.com/information-technology/2012/03/the-pwn-plug-is-a-little-white-box-that-can-hack-your-network/

11. Pourkhomami, P. (2023, September 11). *Types of USB drop attacks & cybersecurity threats with examples*. Managed IT Services & Technology Consulting | OSIbeyond. https://www.osibeyond.com/blog/usb-drop-attacks-cause-cybersecurity-incidents
12. Armasu, L. (2016, September 9). *"USB Killer 2.0" Shows That Most USB-Enabled Devices Are Vulnerable To Power Surge Attacks*. Tom's Hardware. https://www.tomshardware.com/news/usb-killer-2.0-power-surge-attack,32669.html
13. Fruhlinger, J. (2019, May 17). *Malware explained: How to prevent, detect and recover from it*. CSO Online. https://www.csoonline.com/article/565999/what-is-malware-viruses-worms-trojans-and-beyond.html
14. Tung, L. (2020, December 3). *Open source: Almost one in five bugs are planted for malicious purposes*. ZDNET. https://www.zdnet.com/article/open-source-software-how-many-bugs-are-hidden-there-on-purpose/
15. Oladimeji, S., & Kerner, S. M. (2023, November 3). *SolarWinds hack explained: Everything you need to know*. WhatIs; TechTarget. https://www.techtarget.com/whatis/feature/SolarWinds-hack-explained-Everything-you-need-to-know
16. *NSA and CISA red and blue teams share top ten cybersecurity misconfigurations*. (n.d.). Cybersecurity and Infrastructure Security Agency CISA. Retrieved April 23, 2024, from https://www.cisa.gov/news-events/cybersecurity-advisories/aa23-278a
17. Venkat, A. (2023, June 6). *Cloud misconfiguration causes massive data breach at Toyota Motor*. CSO Online. https://www.csoonline.com/article/575483/cloud-misconfiguration-causes-massive-data-breach-at-toyota-motor.html
18. Barker, I. (2023, January 19). *A third of ICS vulnerabilities have no patch available*. BetaNews. https://betanews.com/2023/01/19/a-third-of-ics-vulnerabilities-have-no-patch-available/
19. Wuchner, A. (2022, November 14). *End of Life Software: Risks, Dangers & What to Do Next*. Kiteworks.com. https://www.kiteworks.com/regulatory-compliance/end-life-software-risks
20. *What is a Zero-Day Vulnerability?* (n.d.). Hpe.com. Retrieved April 23, 2024, from https://www.hpe.com/us/en/what-is/zero-day-vulnerability.html
21. Naraine, R. (2022, November 7). *Microsoft: China Flaw Disclosure Law Part of Zero-Day Exploit Surge*. Securityweek.com. https://www.securityweek.com/microsoft-china-flaw-disclosure-law-part-zero-day-exploit-surge
22. Townsend, K. (2024, March 28). *Threat Indicators Show 2024 Is Already Promising to be Worse Than 2023*. Securityweek.com. https://www.securityweek.com/threat-indicators-show-2024-is-already-promising-to-be-worse-than-2023/
23. Greenberg, A. (2016, August 17). The shadow brokers mess is what happens

when the NSA hoards zero-days. *Wired.* https://www.wired.com/2016/08/shadow-brokers-mess-happens-nsa-hoards-zero-days/

5. Assess Your Defenses

1. Clarke, B. (2022, May 4). *Why are security control failures one of the biggest challenges in cybersecurity?* Panaseer; Panaseer Limited. https://panaseer.com/business-blog/security-control-failures-challenge
2. Paganini, P. (2024, April 30). *NCSC: New UK law bans default passwords on smart devices.* Security Affairs. https://securityaffairs.com/162557/laws-and-regulations/ncsc-uk-law-smart-devices.html
3. *MITRE ATT&CK®*. (n.d.). Mitre.org. Retrieved April 23, 2024, from https://attack.mitre.org
4. *CIS Community Defense Model 2.0.* (n.d.). CIS. Retrieved April 23, 2024, from https://www.cisecurity.org/white-papers/cis-community-defense-model-2-0
5. *CIS critical security Controls implementation group 1.* (n.d.). CIS. Retrieved April 23, 2024, from https://www.cisecurity.org/controls/implementation-groups/ig1
6. *CIS Controls implementation group 2.* (n.d.). CIS. Retrieved April 23, 2024, from https://www.cisecurity.org/controls/implementation-groups/ig2
7. *CIS Controls implementation group 3.* (n.d.). CIS. Retrieved April 23, 2024, from https://www.cisecurity.org/controls/implementation-groups/ig3
8. *Washington State Auditor's Office uses CIS Controls to perform effective security audits.* (n.d.). CIS. Retrieved April 23, 2024, from https://www.cisecurity.org/case-study/washington-state-auditors-office-uses-cis-controls-to-perform-effective-security-audits
9. *CIS Controls Self Assessment Tool (CIS CSAT).* (n.d.). CIS. Retrieved April 23, 2024, from https://www.cisecurity.org/controls/cis-controls-self-assessment-tool-cis-csat

7. Present Your Plan

1. *Gramm-Leach-Bliley Act.* (2013, July 19). Federal Trade Commission. https://www.ftc.gov/legal-library/browse/statutes/gramm-leach-bliley-act
2. *FTC Safeguards Rule: What Your Business Needs to Know.* (2022, April 27). Federal Trade Commission. https://www.ftc.gov/business-guidance/resources/ftc-safeguards-rule-what-your-business-needs-know
3. *16 CFR 314.4 -- elements.* (n.d.). Ecfr.gov. Retrieved April 23, 2024, from https://www.ecfr.gov/current/title-16/part-314/section-314.4
4. Wikipedia contributors. (2024, April 5). *Kobayashi Maru.* Wikipedia, The Free Encyclopedia. https://en.wikipedia.org/w/index.php?title=Kobayashi*Maru&oldid=1217377234*

5. Klein, D. (2023, June 21). *CISO stress levels are out of control.* SC Media. https://www.scmagazine.com/perspective/ciso-stress-levels-are-out-of-control
6. *CISOs' churn — CISOs as Chief Incident Scapegoats.* (2023, December 4). 10guards.com. https://10guards.com/en/articles/cisos-churn-cisos-as-chief-incident-scapegoats
7. *SEC Charges SolarWinds and Chief Information Security Officer with Fraud, Internal Control Failures.* (2023, October 30). Sec.gov. https://www.sec.gov/news/press-release/2023-227

8. Execute Your Plan

1. Pratt, M. K. (2023, November 20). *15 benefits of outsourcing your cybersecurity operations.* Security; TechTarget. https://www.techtarget.com/searchsecurity/tip/15-benefits-of-outsourcing-your-cybersecurity-operations

9. Manage Your Risks

1. *Cybersecurity Risk Management, Strategy, Governance, and Incident Disclosure.* (n.d.). Sec.gov. Retrieved April 23, 2024, from https://www.sec.gov/corpfin/secg-cybersecurity

10. Find Your Gaps

1. *Bilge Pump Basics: What is it, and How Should You Use It?* (n.d.). Discover Boating. Retrieved April 23, 2024, from https://www.discoverboating.com/ownership/bilge-pump-basics
2. *CIS critical security Control 7: Continuous vulnerability management.* (n.d.). CIS. Retrieved April 23, 2024, from https://www.cisecurity.org/controls/continuous-vulnerability-management
3. Risto, J. (2023, May 22). *What is CVSS - Common Vulnerability Scoring System.* Sans.org. https://www.sans.org/blog/what-is-cvss
4. *PCI Severity Levels.* (n.d.). Qualys.com. Retrieved April 23, 2024, from https://pci.qualys.com/static/help/merchant/network*scans/pciseverity**levels.htm*
5. *CIS critical security control 18: Penetration testing.* (n.d.). CIS. Retrieved April 23, 2024, from https://www.cisecurity.org/controls/penetration-testing
6. Mukherjee, A. (2023, June 6). *Understanding Social Engineering Penetration Testing.* Threatintelligence.com. https://www.threatintelligence.com/blog/social-engineering-penetration-testing
7. *Physical penetration testing: The most overlooked aspect of security.* (2023,

September 5). ISACA. https://www.isaca.org/resources/white-papers/2023/physical-penetration-testing

8. Jackson, C. (2023, February 23). *Red, blue, and purple teams: Cybersecurity roles explained.* Pluralsight.com. https://www.pluralsight.com/blog/security-professional/red-team-blue-team-purple-team
9. *CIS critical security control 17: Incident response and management.* (n.d.). CIS. Retrieved April 23, 2024, from https://www.cisecurity.org/controls/incident-response-management
10. Lawton, S. (2024, March 4). *4 tabletop exercises every security team should run.* CSO Online. https://www.csoonline.com/article/1311295/4-tabletop-exercises-every-security-team-should-run.html
11. Cranford, J. J. (2021, May 2). *Incident response plan: Frameworks and steps.* Crowdstrike.com; CrowdStrike. https://www.crowdstrike.com/cybersecurity-101/incident-response/incident-response-steps
12. Wlosinski, L. G. (2021, November 2). *Cyberthreat intelligence as a proactive extension to incident response.* ISACA. https://www.isaca.org/resources/isaca-journal/issues/2021/volume-6/cyberthreat-intelligence-as-a-proactive-extension-to-incident-response
13. *CISA tabletop exercise packages.* (n.d.). Cybersecurity and Infrastructure Security Agency CISA. Retrieved April 23, 2024, from https://www.cisa.gov/resources-tools/services/cisa-tabletop-exercise-packages
14. Schroeder, J. (2019, May 28). *Lessons Learned in Incident Response.* Cyber-Sheath. https://cybersheath.com/resources/blog/incident-response-learning-the-lesson-of-lessons-learned/

11. Test Your Defenses

1. *SOC 2® - SOC for Service Organizations: Trust services criteria.* (n.d.). Aicpa-cima.com. Retrieved April 23, 2024, from https://www.aicpa-cima.com/topic/audit-assurance/audit-and-assurance-greater-than-soc-2

Afterword

1. PricewaterhouseCoopers. (n.d.). *Effective cybersecurity starts at the top.* PwC. Retrieved April 23, 2024, from https://www.pwc.com/gx/en/issues/c-suite-insights/the-leadership-agenda/effective-cybersecurity-starts-at-the-top.html
2. The Decision Lab. (n.d.). *Normalcy bias.* The Decision Lab. Retrieved April 23, 2024, from https://thedecisionlab.com/biases/normalcy-bias

ABOUT THE AUTHOR

Michael Lines

Michael Lines is an information security executive with over twenty-five years of experience as a Chief Information Security Officer (CISO), advisory Information Security practice leader, and information security & risk consultant for large, global organizations. Michael writes, blogs, speaks at conferences and webinars, and provides interviews on various information security topics, primarily concerning developing and running effective information security programs and why so many companies continue to suffer security breaches due to ineffective risk management.

Michael is available for consultation regarding cybersecurity risk management, specifically the Heuristic Risk Management approach, and can be reached at https://heuristicsecurity.com/about.

You can subscribe to receive notices of errata and supplemental information on the author's book website at https://heuristicrisk.substack.com.

Made in the USA
Monee, IL
04 November 2026

67385777R10089